PROBLEM-SOLVING CASES IN MICROSOFT® ACCESS™ AND EXCEL®

PROBLEM-SOLVING CASES IN MICROSOFT® ACCESS™ AND EXCEL®

Eighth Annual Edition

Ellen F. Monk

Joseph A. Brady

Australia • Brazil • Japan • Korea • Mexico • Singapore • Spain • United Kingdom • United States

COURSE TECHNOLOGY
CENGAGE Learning™

Problem-Solving Cases in Microsoft® Access™ and Excel®, Eighth Annual Edition
Ellen F. Monk, Joseph A. Brady

Publisher: Joe Sabatino

Senior Acquisitions Editor: Charles McCormick, Jr.

Senior Product Manager: Kate Mason

Development Editor: Dan Seiter

Editorial Assistant: Nora Heink

Marketing Director: Keri Witman

Marketing Manager: Adam Marsh

Senior Marketing Communications Manager: Libby Shipp

Marketing Coordinator: Suellen Ruttkay

Content Project Management: PreMediaGlobal

Media Editor: Chris Valentine

Senior Art Director: Stacy Jenkins Shirley

Cover Designer: Lou Ann Thesing

Cover Image: iStock Photo/©Lise Gagne

Manufacturing Coordinator: Julio Esperas

Compositor: PreMediaGlobal

Library of Congress Control Number: 2010929830

ISBN-13: 978-0-538-48220-2

ISBN-10: 0-538-48220-6

Course Technology
20 Channel Center Street
Boston, MA 02210
USA

Printed in the United States of America
1 2 3 4 5 6 7 16 15 14 13 12 11 10

To Joe, a great colleague, co-author, and friend
EFM

To my co-author, with great appreciation
JAB

BRIEF CONTENTS

Part 4: Decision Support Case Using Basic Excel Functionality

Part 5: Integration Cases Using Access and Excel

Part 6: Advanced Excel Skills

Part 7: Presentation Skills

PREFACE

For two decades, we have taught MIS courses at the University of Delaware. From the start, we wanted to use good computer-based case studies for the database and decision-support portions of our courses.

We could not find a casebook that met our needs! This surprised us because we thought our requirements were not unreasonable. First, we wanted cases that asked students to think about real-world business situations. Second, we wanted cases that provided students with hands-on experience, using the kind of software that they had learned to use in their computer literacy courses—and that they would later use in business. Third, we wanted cases that would strengthen students' ability to analyze a problem, examine alternative solutions, and implement a solution using software. Undeterred by the lack of casebooks, we wrote our own cases, and Course Technology, part of Cengage Learning, published them.

This is the eighth casebook we have written for Course Technology. The cases are all new and the tutorials are updated.

As with our prior casebooks, we include tutorials that prepare students for the cases, which are challenging but doable. Most of the cases are organized in a way that helps students think about the logic of each case's business problem and then about how to use the software to solve the business problem. The cases fit well in an undergraduate MIS course, an MBA Information Systems course, or a Computer Science course devoted to business-oriented programming.

BOOK ORGANIZATION

The book is organized into seven parts:

- Database Cases Using Access
- Decision Support Cases Using the Excel Scenario Manager
- Decision Support Cases Using the Excel Solver
- Decision Support Case Using Basic Excel Functionality
- Integration Cases Using Access and Excel
- Advanced Excel Skills
- Presentation Skills

Part 1 begins with two tutorials that prepare students for the Access case studies. Parts 2 and 3 each begin with a tutorial that prepares students for the Excel case studies. All four tutorials provide students with hands-on practice in using the software's more advanced features—the kind of support that other books about Access and Excel do not give to students. Part 4 asks students to use Excel's basic functionality for decision support. Part 5 challenges students to use both Access and Excel to find a solution to a business problem. Part 6 is a set of short tutorials on the advanced skills that students need to complete some of the Excel cases. Part 7 is a tutorial that hones students' skills in creating and delivering an oral presentation to business managers. The next section explores each of these parts in more depth.

Part 1: Database Cases Using Access

This section begins with two tutorials and then presents five case studies.

Tutorial A: Database Design

This tutorial helps students to understand how to set up tables to create a database, without requiring students to learn formal analysis and design methods, such as data normalization.

Tutorial B: Microsoft Access Tutorial

The second tutorial teaches students the more advanced features of Access queries and reports—features that students will need to know to complete the cases.

Cases 1–5

Five database cases follow Tutorials A and B. The students' job is to implement each case's database in Access so form, query, navigation pane, and report outputs can help management. The first case is an easier "warm-up" case. The next four cases require a more demanding database design and implementation effort.

Part 2: Decision Support Cases Using Excel Scenario Manager

This section has one tutorial and two decision support cases that require the use of the Excel Scenario Manager.

Tutorial C: Building a Decision Support System in Excel

This section begins with a tutorial using Excel for decision support and spreadsheet design. Fundamental spreadsheet design concepts are taught. Instruction on the Scenario Manager, which can be used to organize the output of many "what-if" scenarios, is emphasized.

Cases 6–7

These two cases can be done with or without the Scenario Manager (although the Scenario Manager is nicely suited to them). In each case, students must use Excel to model two or more solutions to a problem. Students then use the outputs of the model to identify and document the preferred solution via a memorandum and, if assigned to do so, an oral presentation.

Part 3: Decision Support Cases Using the Excel Solver

This section has one tutorial and two decision support cases requiring the use of Excel Solver.

Tutorial D: Building a Decision Support System Using the Excel Solver

This section begins with a tutorial about using the Solver, which is a decision support tool for solving optimization problems.

Cases 8–9

Once again, in each case, students use Excel to analyze alternatives and identify and document the preferred solution.

Part 4: Decision Support Case Using Basic Excel Functionality

Case 10

The cases continue with one case that uses basic Excel functionality. (In other words, the case does not require the Scenario Manager or the Solver.) Excel is used to test students' analytical skills in "what-if" analyses.

Part 5: Integration Cases Using Access and Excel

Cases 11 and 12

These cases integrate Access and Excel. These cases are included because of a trend toward sharing data among multiple software packages to solve problems.

Part 6: Advanced Excel Skills

This part has one tutorial focused on advanced techniques in Excel.

Tutorial E: Guidance for Excel Cases

A number of cases in this book require the use of some advanced techniques in Excel. Techniques for using data tables and pivot tables are explained in Tutorial E rather than in the cases themselves.

Part 7: Presentation Skills

Tutorial F: Giving an Oral Presentation

Each case includes an optional assignment that gives students practice in making a presentation to management on the results of their case analysis. This section gives advice on how to create oral presentations. It also has technical information on charting, a technique that might be useful in case analyses or as support for presentations. This tutorial will help students to organize their recommendations, to present their solutions in both words and graphics, and to answer questions from the audience. For larger classes, instructors may want to have students work in teams to create and deliver their presentations—which would model the "team" approach used by many corporations.

To view and access additional cases, instructors should see the "Hall of Fame" note in the *Using the Cases* section shown below.

INDIVIDUAL CASE DESIGN

The format of the cases uses the following template:

- Each case begins with a *Preview* and an overview of the tasks.
- The next section, *Preparation*, tells students what they need to do or know to complete the case successfully. (Of course, our tutorials also prepare students for the cases!)
- The third section, *Background*, provides the business context that frames the case. The background of each case models situations that require the kinds of thinking and analysis that students will need in the business world.
- The Background sections are followed by the *Assignment* sections, which are generally organized in a way that helps students to develop their analyses.
- The last section, *Deliverables*, lists what students must hand in: printouts, a memorandum, a presentation, and files on disk. The list is similar to the kind of deliverables that a business manager might demand.

USING THE CASES

We have successfully used cases like these in our undergraduate MIS courses. We usually begin the semester with Access database instruction. We assign the Access database tutorials and then a case to each student. Then, for Excel decision support system instruction, we do the same thing: assign a tutorial and then a case.

Some instructors have expressed an interest in having access to extra cases, especially in the second semester of a school year. For example: "I assigned the integration case in the fall, and need another one for the spring." To meet this need, we have set up an online "Hall of Fame" that features some of our favorite cases from prior editions. This password-protected Hall of Fame is available to instructors on the Cengage Learning Web site. Go to *www.cengage.com/coursetechnology* and search for this textbook by title, author, or ISBN. Note that the cases are in MS Office 2003 format, but MS Office 2007 and Office 2010 will read and translate them easily.

TECHNICAL INFORMATION

This textbook was tested for quality assurance using the Windows 7 operating system, Microsoft Access 2007, and Microsoft Excel 2007.

Data Files and Solution Files

We have created "starter" data files for the Excel cases, so students need not spend time typing in the spreadsheet skeleton. Cases 11 and 12 also require students to load an Access database file. All these files are on the Cengage Learning Web site, which is available to both students and instructors. Go to *www.cengage.com/coursetechnology* and search for this textbook by title, author, or ISBN. You are granted a license to copy the data files to any computer or computer network used by people who have purchased this textbook.

Solutions to the material in the text are available to instructors at *www.cengage.com/coursetechnology*. Search for this textbook by title, author, or ISBN. The solutions are password protected.

Instructor's Manual

An Instructor's Manual is available to accompany this text. To help instructors successfully use the textbook, the Instructor's Manual contains tools and information such as a Sample Syllabus, Teaching Tips, and Grading Guidelines. Instructors should go to *www.cengage.com/coursetechnology* and search for this textbook by title, author, or ISBN. The Instructor's Manual is password protected.

ACKNOWLEDGEMENTS

We would like to give many thanks to the team at Cengage Learning, including our Developmental Editor, Dan Seiter; Product Manager, Kate Hennessy Mason; and our Content Project Manager, Karunakaran Gunasekaran. As always, we acknowledge our students' diligent work.

PART 1

DATABASE CASES USING ACCESS

TUTORIAL **A**

DATABASE DESIGN

This tutorial has three sections. The first section briefly reviews basic database terminology. The second section teaches database design. The third section features a practice database design problem.

REVIEW OF TERMINOLOGY

You will begin by reviewing some basic terms that will be used throughout this textbook. In Access, a **database** is a group of related objects that are saved in one file. An Access **object** can be a table, a form, a query, or a report. You can identify an Access database file by its suffix .accdb.

A **table** consists of data that is arrayed in rows and columns. A **row** of data is called a **record**. A **column** of data is called a **field**. Thus, a record is a set of related fields. The fields in a table should be related to one another in some way. For example, a company might want to keep its employee data together by creating a database table called EMPLOYEE. That table would contain data fields about employees, such as their names and addresses. It would not have data fields about the company's customers; that data would go in a CUSTOMER table.

A field's values have a **data type** that is declared when a table is defined. That way, when data is entered into the database, the software knows how to interpret each entry. The data types found in Access include the following:

- "Text" for words
- "Integer" for whole numbers
- "Double" for numbers that have a decimal value
- "Currency" for numbers that should be treated as dollars and cents
- "Yes/No" for variables that have only two values (1-0, on/off, yes/no, true/false)
- "Date/Time" for variables that are dates or times

Each database table should have a **primary key** field, a field in which each record has a *unique* value. For example, in an EMPLOYEE table, a field called SSN (for Social Security number) could serve as a primary key, because each record's SSN value would be different from every other record's SSN value. Sometimes a table does not have a single field whose values are all different. In that case, two or more fields are combined into a **compound primary key**. The combination of the fields' values is unique.

Database tables should be logically related to one another. For example, suppose a company has an EMPLOYEE table with fields for SSN, Name, Address, and Telephone Number. For payroll purposes, the company has an HOURS WORKED table with a field that summarizes Labor Hours for individual employees. The relationship between the EMPLOYEE table and the HOURS WORKED table needs to be established in the database so you can tell which employees worked which hours. That is done by including the primary key field from the EMPLOYEE table (SSN) as a field in the HOURS WORKED table. In the HOURS WORKED table, the SSN field is then called a **foreign key**.

In Access, data can be entered directly into a table or it can be entered into a form, which then inserts the data into a table. A **form** is a database object that is created from an existing table to make the process of entering data more user-friendly.

A **query** is the database equivalent of a question that is posed about data in a table (or tables). For example, suppose a manager wants to know the names of employees who have worked for the company for more than five years. A query could be designed in such a way that it interrogates the EMPLOYEE table to search for the information. The query would be run, and its output would answer the question.

Because a query may need to pull data from more than one table, queries can be designed to interrogate more than one table at a time. In that case, the tables must be connected by a **join** operation, which links tables on the values in a field that they have in common. The common field acts as a "hinge" for the joined tables; when the query is run, the query generator treats the joined tables as one large table.

In Access, queries that answer a question are called select queries, because they select relevant data from the database records. Queries also can be designed to change data in records, add a record to the end of a table, or delete entire records from a table. Those are called **update**, **append**, and **delete** queries, respectively.

Access has a **report** generator that can be used to format a table's data or a query's output.

DATABASE DESIGN

Designing a database involves determining which tables need to be in the database and creating the fields that need to be in each table. This section begins with an introduction to key database design concepts, then discusses database design rules. Those rules are a series of steps you should use when building a database. First, the following key concepts are defined:

- Entities
- Relationships
- Attributes

Database Design Concepts

Computer scientists have highly formalized ways of documenting a database's logic, but learning their notations and mechanics can be time-consuming and difficult. In fact, doing so usually takes a good portion of a systems analysis and design course. This tutorial will teach you database design by emphasizing practical business knowledge, and the approach should enable you to design serviceable databases. Your instructor may add more formal techniques.

A database models the logic of an organization's operation, so your first task is to understand that operation. You do that by talking to managers and workers, by making observations, and/or by looking at business documents such as sales records. Your goal is to identify the business's "entities" (sometimes called *objects*). An **entity** is some thing or some event that the database will contain. Every entity has characteristics, called **attributes**, and a **relationship** (or relationships) to other entities. Take a closer look.

Entities

As previously mentioned, an entity is a tangible thing or an event. The reason for identifying entities is that *an entity eventually becomes a table in the database*. Entities that are things are easy to identify. For example, consider a video store. The database for the video store would probably need to contain the names of DVDs and the names of customers who rent them, so you would have one entity named VIDEO and another named CUSTOMER.

In contrast, entities that are events can be more difficult to identify, probably because although events cannot be seen, they are no less real. In the video store example, one event would be VIDEO RENTAL and another event would be HOURS WORKED by employees.

In general, your analysis of an organization's operations can be made easier by knowing that organizations usually have certain physical entities such as these:

- Employees
- Customers
- Inventory (products or services)
- Suppliers

Thus, the database for most organizations would have a table for each of those entities. Your analysis also can be made easier by knowing that organizations engage in transactions internally (within the company) and externally (with the outside world). Those transactions are the subject of any accounting course, but most people understand them from events that occur in daily life. Consider the following examples:

- Organizations generate revenue from sales or interest earned. Revenue-generating transactions include event entities called SALES and INTEREST.
- Organizations incur expenses from paying hourly employees and purchasing materials from suppliers. HOURS WORKED and PURCHASES are event entities in the databases of most organizations.

Thus, identifying entities is a matter of observing what happens in an organization. Your powers of observation are aided by knowing what entities exist in the databases of most organizations.

Relationships

As a database analyst building a database, you should consider the relationship of each entity to other entities. For each entity, you should ask, "What is the relationship, if any, of this entity to every other entity identified?" Relationships can be expressed in English. For example, suppose a college's database has entities for STUDENT (containing data about each student), COURSE (containing data about each course), and SECTION (containing data about each section). A relationship between STUDENT and SECTION would be expressed as "Students enroll in sections."

An analyst also must consider the **cardinality** of any relationship. Cardinality can be one-to-one, one-to-many, or many-to-many. Those relationships are summarized as follows:

- In a one-to-one relationship, one instance of the first entity is related to just one instance of the second entity.
- In a one-to-many relationship, one instance of the first entity is related to many instances of the second entity, but each instance of the second entity is related to only one instance of the first entity.
- In a many-to-many relationship, one instance of the first entity is related to many instances of the second entity and one instance of the second entity is related to many instances of the first entity.

For a more concrete understanding of cardinality, consider again the college database with the STUDENT, COURSE, and SECTION entities. The university catalog shows that a course such as Accounting 101 can have more than one section: 01, 02, 03, 04, etc. Thus, the following relationships can be observed:

- The relationship between the entities COURSE and SECTION is one-to-many. Each course has many sections, but each section is associated with just one course.
- The relationship between STUDENT and SECTION is many-to-many. Each student can be in more than one section, because each student can take more than one course. Also, each section has more than one student.

Thinking about relationships and their cardinalities may seem tedious to you. But as you work through the cases in this text, you will see that this type of analysis and the knowledge it yields can be very valuable in designing databases. In the case of many-to-many relationships, you should determine the database tables a given database needs; in the case of one-to-many relationships, you should decide which fields the database's tables need to share.

Attributes

An attribute is a characteristic of an entity. The reason you identify attributes of an entity is because *attributes become a table's fields*. If an entity can be thought of as a noun, an attribute can be thought of as an adjective describing the noun. Continuing with the college database example, consider the STUDENT entity. Students have names. Thus, Last Name would be an attribute of the entity called STUDENT and, therefore, a field in the STUDENT table. First Name would be an attribute as well. The STUDENT entity also would have an Address attribute as another field, along with Phone Number and whatever other fields were used.

Sometimes it can be difficult to tell the difference between an attribute and an entity. One good way to differentiate them is to ask whether more than one attribute is possible for each entity. If more than one instance is possible and you do not know in advance how many there will be, then it's an entity. For example, assume a student could have two (but no more than two) Addresses—one for home and one for college. You could specify attributes Address 1 and Address 2. Now consider what would happen if the number of student addresses could not be stipulated in advance, meaning all addresses had to be recorded. In that case, you would not know how many fields to set aside in the STUDENT table for addresses. Therefore, you would need a separate STUDENT ADDRESSES table (entity) that would show any number of addresses for a given student.

DATABASE DESIGN RULES

As described previously, your first task in database design is to understand the logic of the business situation. Once you understand that, you are ready to build a database for the requirements of the situation. To create a context for learning about database design, look at a hypothetical business operation and its database needs.

Example: The Talent Agency

Suppose you have been asked to build a database for a talent agency. The agency books bands into nightclubs. The agent needs a database to keep track of the agency's transactions and to answer day-to-day questions. Many questions arise in the running of this business. For example, a club manager often wants to know which bands are available on a certain date at a certain time or wants to know the agent's fee for a certain band. Similarly, the agent may want to see a list of all band members and the instrument each person plays or a list of all bands having three members.

Suppose you have talked to the agent and have observed the agency's business operation. You conclude that your database needs to reflect the following facts:

1. A booking is an event in which a certain band plays in a particular club on a particular date, starting at a certain time, ending at a certain time, and performing for a specific fee. A band can play more than once a day. The Heartbreakers, for example, could play at the East End Cafe in the afternoon and then at the West End Cafe that same night. For each booking, the club pays the talent agent. The agent keeps a 5 percent fee and then gives the remainder of the payment to the band.
2. Each band has at least two members and an unlimited maximum number of members. The agent notes a telephone number of just one band member, which is used as the band's contact number. No two bands have the same name or telephone number.
3. No members of any of the bands have the same name. For example, if there is a Sally Smith in one band, there is no Sally Smith in another band.
4. The agent keeps track of just one instrument that each band member plays. For this record keeping purpose, "vocals" are considered an instrument.
5. Each band has a desired fee. For example, the Lightmetal band might want $700 per booking and would expect the agent to try to get at least that amount.
6. Each nightclub has a name, an address, and a contact person. That person has a telephone number that the agent uses to contact the club. No two clubs have the same name, contact person, or telephone number. Each club has a target fee. The contact person will try to get the agent to accept that amount for a band's appearance.
7. Some clubs feed the band members for free; others do not.

Before continuing with this tutorial, you might try to design the agency's database on your own. Ask yourself, what are the entities? Recall that databases usually have CUSTOMER, EMPLOYEE, and INVENTORY entities as well as an entity for the revenue-generating transaction event. Each entity becomes a table in the database. What are the relationships between entities? For each entity, what are its attributes? For each table, what is the primary key?

Six Database Design Rules

Assume you have gathered information about the business situation in the talent agency example. Now you want to identify the tables required for the database and the fields needed in each table. To do that, observe the following six rules:

Rule 1: You do not need a table for the business. The database represents the entire business. Thus, in the example, Agent and Agency are not entities.

Rule 2: Identify the entities in the business description. Look for the things and events that the database must contain. Those become tables in the database. Typically, certain entities are represented. In the talent agency example, you should be able to observe these entities:

- *Things*: The product (inventory for sale) is Band. The customer is Club.
- *Events*: The revenue-generating transaction is Bookings.

You might ask yourself, Is there an EMPLOYEE entity? Isn't INSTRUMENT an entity? Those issues will be discussed as the rules are explained.

Rule 3: Look for relationships between the entities. Look for one-to-many relationships between entities. The relationship between those entities must be established in the tables, and that is done by using a foreign key. Those mechanics are explained in the next rule's discussion of the relationship between Band and Band Member.

Look for many-to-many relationships between entities. In each of those relationships, there is the need for a third entity that associates the two entities in the relationship. Recall from the college database scenario the many-to-many relationship example that involved STUDENT and SECTION entities. To show the ENROLLMENT of specific students in specific sections, a third table needs to be created. The mechanics of doing that are described in the next rule in the discussion of the relationship between BAND and CLUB.

Rule 4: Look for attributes of each entity and designate a primary key. As previously mentioned, you should think of the entities in your database as nouns. You should then create a list of adjectives that describe those nouns. Those adjectives are the attributes that will become the table's fields. After you have identified fields for each table, you should check to see if a field has unique values. If one exists, designate it as the primary key field; otherwise, designate a compound primary key.

Returning to the talent agency example, the attributes, or fields, of the BAND entity are Band Name, Band Phone Number, and Desired Fee. No two bands have the same names, so the primary key field can be Band Name. Figure A-1 shows the BAND table and its fields: Band Name, Band Phone Number, and Desired Fee; the data type of each field also is shown.

BAND	
Field	**Data Type**
Band Name (primary key)	Text
Band Phone Number	Text
Desired Fee	Currency

FIGURE A-1 The BAND table and its fields

Two BAND records are shown in Figure A-2.

Band Name (primary key)	Band Phone Number	Desired Fee
Heartbreakers	981 831 1765	$800
Lightmetal	981 831 2000	$700

FIGURE A-2 Records in the BAND table

If two bands could have the same name (that is, if uniqueness in band names wasn't assumed), Band Name would not be a good primary key and some other unique identifier would be needed. Those situations are common. Most businesses have many types of inventory, and duplicate names are possible. The typical solution is to assign a number to each product to be used as the primary key field. For example, a college could have more than one faculty member with the same name, so each faculty member would be assigned an employee identification number (EIN). Similarly, banks assign a personal identification number (PIN) for each depositor. Each automobile that a car manufacturer makes gets a unique Vehicle Identification Number (VIN). Most businesses assign a number to each sale, called an invoice number. (The next time you buy something at a grocery store, note the number on your receipt. It will be different from the number the next person in line sees on his or her receipt.)

At this point, you might be wondering why Band Member would not be an attribute of BAND. The answer is that although you must record each band member, you do not know in advance how many members will be in each band. Therefore, you do not know how many fields to allocate to the BAND table for members. Another way to think about Band Member(s) is that they are, in effect, the agency's employees. Databases for organizations usually have an EMPLOYEE entity. Therefore, you should create a BAND MEMBER table with the attributes Member Name, Band Name, Instrument, and Phone. The BAND MEMBER table and its fields are shown in Figure A-3.

BAND MEMBER	
Field Name	**Data Type**
Member Name (primary key)	Text
Band Name (foreign key)	Text
Instrument	Text
Phone	Text

FIGURE A-3 The BAND MEMBER table and its fields

Note in Figure A-3 that the phone number is classified as a text data type. The data type for such "numbers" is text—and not number—because the values will not be used in an arithmetic computation. The benefit is that text data type values take up fewer bytes than numerical or currency data type values; therefore, the file uses less storage space. That rule of using text data types would also hold for values such as zip codes, Social Security numbers, etc.

Five records in the BAND MEMBER table are shown in Figure A-4.

Member Name (primary key)	Band Name	Instrument	Phone
Pete Goff	Heartbreakers	Guitar	981 444 1111
Joe Goff	Heartbreakers	Vocals	981 444 1234
Sue Smith	Heartbreakers	Keyboard	981 555 1199
Joe Jackson	Lightmetal	Sax	981 888 1654
Sue Hoopes	Lightmetal	Piano	981 888 1765

FIGURE A-4 Records in the BAND MEMBER table

Instrument can be included as a field in the BAND MEMBER table because the agent records only one instrument for each band member. Thus, the instrument can be thought of as a way to describe a band member, much like the phone number is part of the description. Member Name can be the primary key because of the assumption (albeit arbitrary) that no two members in any band have the same name. Alternatively, Phone could be the primary key, assuming no two members share a telephone. Or a band member ID number could be assigned to each person in each band, which would create a unique identifier for each band member that the agency handled.

You might ask why Band Name is included in the BAND MEMBER table. The commonsense reason is that you did not include the Member Name in the BAND table. You must relate bands and members somewhere, and the BAND table is the place to do it.

Another way to think about this involves the cardinality of the relationship between BAND and BAND MEMBER. It is a one-to-many relationship: one band has many members, but each member is in just one band. You establish that kind of relationship in the database by using the primary key field of one table as a foreign key in the other table. In BAND MEMBER, the foreign key Band Name is used to establish the relationship between the member and his or her band.

The attributes of the entity CLUB are Club Name, Address, Contact Name, Club Phone Number, Preferred Fee, and Feed Band?. The table called CLUB can define the CLUB entity, as shown in Figure A-5.

CLUB	
Field Name	**Data Type**
Club Name (primary key)	Text
Address	Text
Contact Name	Text
Club Phone Number	Text
Preferred Fee	Currency
Feed Band?	Yes/No

FIGURE A-5 The CLUB table and its fields

Two records in the CLUB table are shown in Figure A-6.

Club Name (primary key)	Address	Contact Name	Club Phone Number	Preferred Fee	Feed Band?
East End	1 Duce St.	Al Pots	981 444 8877	$600	Yes
West End	99 Duce St.	Val Dots	981 555 0011	$650	No

FIGURE A-6 Records in the CLUB table

You might wonder why Bands Booked into Club (or some such field name) is not an attribute of the CLUB table. There are two reasons. First, because you do not know in advance how many bookings a club will have, the value cannot be an attribute. Second, BOOKINGS is the agency's revenue-generating transaction, an event entity, and you need a table for that business transaction. Consider the booking transaction next.

You know that the talent agent books a certain band into a certain club on a certain date for a certain fee, starting and ending at a certain time. From that information, you can see that the attributes of the BOOKINGS entity are Band Name, Club Name, Date, Start Time, End Time, and Fee. The BOOKINGS table and its fields are shown in Figure A-7.

BOOKINGS	
Field Name	**Data Type**
Band Name	Text
Club Name	Text
Date	Date/Time
Start Time	Date/Time
End Time	Date/Time
Fee	Currency

FIGURE A-7 The BOOKINGS table and its fields—and no designation of a primary key

Some records in the BOOKINGS table are shown in Figure A-8.

Band Name	Club Name	Date	Start Time	End Time	Fee
Heartbreakers	East End	11/21/10	19:00	23:30	$800
Heartbreakers	East End	11/22/10	19:00	23:30	$750
Heartbreakers	West End	11/28/10	13:00	18:00	$500
Lightmetal	East End	11/21/10	13:00	18:00	$700
Lightmetal	West End	11/22/10	13:00	18:00	$750

FIGURE A-8 Records in the BOOKINGS table

Note that no single field is guaranteed to have unique values because each band is likely to be booked many times and each club used many times. Further, each date and time also can appear more than once. Thus, no one field can be the primary key.

If a table does not have a single primary key field, you can make a compound primary key whose field values, when taken together, will be unique. Because a band can be in only one place at a time, one possible solution is to create a compound key consisting of the fields Band Name, Date, and Start Time. An alternative solution is to create a compound primary key consisting of the fields Club Name, Date, and Start Time.

A way to avoid having a compound key is to create a field called Booking Number. Each booking would have its own unique number, similar to an invoice number.

Here is another way to think about this event entity: Over time, a band plays in many clubs and each club hires many bands. Thus, the BAND-to-CLUB relationship is a many-to-many relationship. Such relationships signal the need for a table between the two entities in the relationship. Here you would need the BOOKINGS table, which associates the BAND and CLUB tables. An associative table is implemented by including the primary keys from the two tables that are associated. In this case, the primary keys from the BAND and CLUB tables are included as foreign keys in the BOOKINGS table.

Rule 5: Avoid data redundancy. You should not include extra (redundant) fields in a table. Doing so takes up extra disk space and leads to data entry errors because the same value must be entered in multiple tables, increasing the chance of a keystroke error. In large databases, keeping track of multiple instances of the same data is nearly impossible and contradictory data entries become a problem.

Consider this example: Why wouldn't Club Phone Number be in the BOOKINGS table as a field? After all, the agent might have to call about some last-minute change for a booking and could quickly look up the number in the BOOKINGS table. Assume the BOOKINGS table had Booking Number as the primary key and Club Phone Number as a field. Figure A-9 shows the BOOKINGS table with the additional field.

BOOKINGS	
Field Name	**Data Type**
Booking Number (primary key)	Text
Band Name	Text
Club Name	Text
Club Phone Number	Text
Date	Date/Time
Start Time	Date/Time
End Time	Date/Time
Fee	Currency

FIGURE A-9 The BOOKINGS table with an unnecessary field—Club Phone Number

The fields Date, Start Time, End Time, and Fee logically depend on the Booking Number primary key—they help define the booking. Band Name and Club Name are foreign keys and are needed to establish the relationship between the tables BAND, CLUB, and BOOKINGS. But what about Club Phone Number? It is not defined by the Booking Number. It is defined by Club Name—*that is, it's a function of the club, not of the booking*. Thus, the Club Phone Number field does not belong in the BOOKINGS table. It's already in the CLUB table; if the agent needs the Club Phone Number field, he or she can look it up there.

Perhaps you can see the practical data entry problem of including Club Phone Number in BOOKINGS. Suppose a club changed its contact phone number. The agent can easily change the number one time, in CLUB. But now the agent would need to remember the names of all the other tables that had that field and change the values there too. Of course, with a small database, that might not be difficult. But in large databases, having many redundant fields in many tables makes this sort of maintenance very difficult, which means that redundant data is often incorrect.

You might object, saying, "What about all of those foreign keys? Aren't they redundant?" In a sense, they are. But they are needed to establish the relationship between one entity and another, as discussed previously.

Rule 6: Do not include a field if it can be calculated from other fields. A **calculated field** is made using the query generator. Thus, the agent's fee is not included in the BOOKINGS table, because it can be calculated by query (here, 5 percent times the booking fee).

PRACTICE DATABASE DESIGN PROBLEM

Imagine this scenario: Your town has a library that wants to keep track of its business in a database, and you have been called in to build the database. You talk to the town librarian, review the old paper-based records, and watch people use the library for a few days. You learn these things about the library:

1. Anyone who lives in the town can get a library card if he or she asks for one. The library considers each person who gets a card a "member" of the library.
2. The librarian wants to be able to contact members by telephone and by mail. She calls members when their books are overdue or when requested materials become available. She likes to mail a thank-you note to each patron on his or her anniversary of becoming a member of the library. Without a database, contacting members can be difficult to do efficiently; for example, there could be more than one member by the name of Martha Jones. Also, a parent and a child can have the same first and last name, live at the same address, and share a phone.
3. The librarian tries to keep track of each member's reading interests. When new books come in, the librarian alerts members whose interests match those books. For example, long-time member Sue Doaks is interested in reading Western novels, growing orchids, and baking bread. There must be some way to match her interests with available books. One complication is that

although the librarian wants to track all of a member's reading interests, she wants to classify each book as being in just one category of interest. For example, the classic gardening book *Orchids of France* would be classified as a book about orchids or a book about France, but not both.

4. The library stocks many books. Each book has a title and any number of authors. Also, there could conceivably be more than one book in the library titled *History of the United States*. Similarly, there could be more than one author with the same name.
5. A writer could be the author of more than one book.
6. A book could be checked out repeatedly as time goes on. For example, *Orchids of France* could be checked out by one member in March, another member in July, and yet another member in September.
7. The library must be able to identify whether a book is checked out.
8. A member can check out any number of books in a visit. It's also conceivable that a member could visit the library more than once a day to check out books—some members do.
9. All books that are checked out are due back in two weeks, with no exceptions. The late fee is 50 cents per day. The librarian would like to have an automated way of generating an overdue book list each day so she can telephone the miscreants.
10. The library has a number of employees. Each employee has a job title. The librarian is paid a salary, but other employees are paid by the hour. Employees clock in and out each day. Assume all employees work only one shift per day and all are paid weekly. Pay is deposited directly into an employee's checking account—no checks are hand-delivered. The database needs to include the librarian and all other employees.

Design the library's database, following the rules set forth in this tutorial. Your instructor will specify the format of your work. Here are a few hints in the form of questions:

- A book can have more than one author. An author can write more than one book. How would you describe the relationship between books and authors?
- The library lends books for free, of course. If you were to think of checking out a book as a sale transaction for zero revenue, how would you handle the library's revenue-generating event?
- A member can borrow any number of books at a checkout. A book can be checked out more than once. How would you describe the relationship between checkouts and books?

MICROSOFT ACCESS TUTORIAL

Microsoft Access is a relational database package that runs on the Microsoft Windows operating system. This tutorial was prepared using Access 2007.

Before using this tutorial, you should know the fundamentals of Microsoft Access and know how to use Windows. This tutorial teaches you some advanced Access skills you'll need to do database case studies. The tutorial concludes with a discussion of common Access problems and how to solve them.

A word of caution: Always observe proper file-saving and closing procedures. Use these steps to exit Access: (1) Office button—Close Database, then (2) X-Exit Access button. Or, you can simply click the X-Exit Access button, which takes you back to Windows. Always end your work with these steps. If you remove your disk, CD, or other portable storage device when database forms and tables appear on the screen, you will lose your work.

To begin this tutorial, you will create a new database called Employee.

AT THE KEYBOARD

Open a new database. On the Getting Started with Microsoft Office Access page, under the heading New Blank Database, click Blank Database. Name the database Employee. Click the file folder next to the filename to browse for the desired folder. Otherwise, your file will be saved automatically in My Documents.

Your opening screen should resemble the screen shown in Figure B-1.

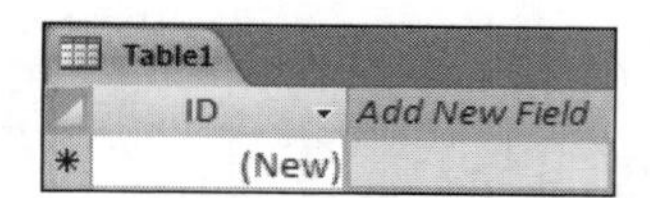

FIGURE B-1 Entering data in Datasheet view

When you create a table, Access opens it in Datasheet view by default. Because you will use Design view to build your tables, close the new table by clicking the *X* in the upper-right corner of the table window that corresponds to Close Table I. You are now on the Home tab in the Database window of Access, as shown in Figure B-2. From this screen, you can create or change objects.

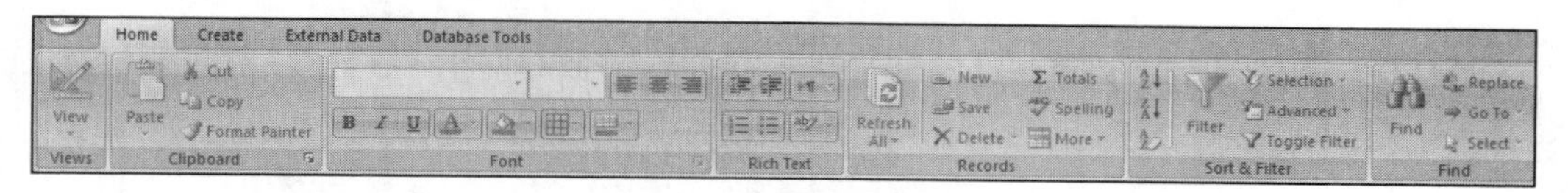

FIGURE B-2 The Database window Home tab in Access

CREATING TABLES

Your database will contain data about employees, their wage rates, and the hours they worked.

Defining Tables

In the Database window, build three new tables using the instructions that follow.

AT THE KEYBOARD

(1) Define a table called EMPLOYEE

This table contains permanent data about employees. To create it, choose the Create tab; then, in the Tables group, click Table design. The table's fields are Last Name, First Name, Employee ID, Street Address, City, State, Zip, Date Hired, and US Citizen. The Employee ID field is the primary key field. Change the lengths of text fields from the default 255 spaces to more appropriate lengths; for example, the Last Name field might be 30 spaces and the Zip field might be 10 spaces. Your completed definition should resemble the one shown in Figure B-3.

Table1

Field Name	Data Type	Description
Last Name	Text	
First Name	Text	
Employee ID	Text	
Street Address	Text	
City	Text	
State	Text	
Zip	Text	
Date Hired	Date/Time	
US Citizen	Yes/No	

FIGURE B-3 Fields in the EMPLOYEE table

When you finish, choose Office button—Save, and then enter the name you want for the table (here, EMPLOYEE). Make sure to specify the name of the *table*, not the database itself. (Here, it is a coincidence that the EMPLOYEE table has the same name as its database file.) Close the table by clicking the Close button (X) that corresponds to the EMPLOYEE table.

(2) Define a table called WAGE DATA

This table contains permanent data about employees and their wage rates. The table's fields are Employee ID, Wage Rate, and Salaried. The Employee ID field is the primary key field. Use the data types shown in Figure B-4. Your definition should resemble the one shown in Figure B-4.

Table1

Field Name	Data Type	Description
Employee ID	Text	
Wage Rate	Currency	
Salaried	Yes/No	

FIGURE B-4 Fields in the WAGE DATA table

Use Office button—Save to save the table definition. Name the table WAGE DATA.

(3) Define a table called HOURS WORKED

The purpose of this table is to record the number of hours that employees work each week during the year. The table's fields are Employee ID (text), Week # (number—long integer), and Hours (number—double). The Employee ID and Week # are the compound keys.

In the following example, the employee with ID number 08965 worked 40 hours in Week 1 of the year and 52 hours in Week 2.

Employee ID	Week#	Hours
08965	1	40
08965	2	52

Note that no single field can be the primary key field, because 08965 is an entry for each week. In other words, if this employee works each week of the year, 52 records will have the Employee ID value at the end of the year. Thus, Employee ID values will not distinguish records. In addition, no other single field can distinguish these records either because other employees will have worked during the same week number and some employees will have worked the same number of hours. For example, 40 hours—which corresponds to a full-time workweek—would be a common entry for many weeks.

All of this presents a problem because in Access, a table must have a primary key field. What is the solution? Use a compound primary key; that is, use values from more than one field to create a combined field that will distinguish records. The best compound key to use for the current example consists of the Employee ID field and the Week # field. Why? Because as each person works each week, the week passes. That means, for example, that there is only *one* combination of Employee ID 08965 and Week # 1. Because those values *can occur in only one record*, the combination distinguishes that record from all others.

The first step of setting a compound key is to highlight the fields in the key. Those fields must appear one after the other in the table definition screen. (Plan ahead for that format.) As an alternative, you can highlight one field, hold down the Control key, and highlight the next field.

AT THE KEYBOARD

For the HOURS WORKED table, click the first field's left prefix area (known as the row selector), hold down the mouse button, and drag down to highlight names of all fields in the compound primary key. Your screen should resemble the one shown in Figure B-5.

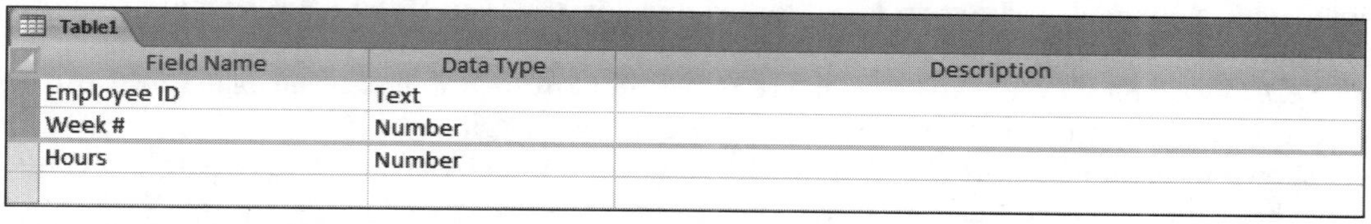

Table1

Field Name	Data Type	Description
Employee ID	Text	
Week #	Number	
Hours	Number	

FIGURE B-5 Selecting fields for the compound primary key for the HOURS WORKED table

Now click the Key icon. Your screen should resemble the one shown in Figure B-6.

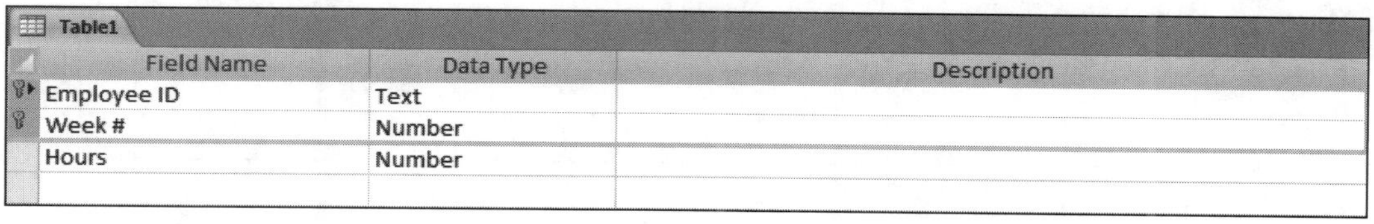

Table1

Field Name	Data Type	Description
Employee ID	Text	
Week #	Number	
Hours	Number	

FIGURE B-6 The compound primary key for the HOURS WORKED table

You have created the compound primary key and finished defining the table. Use Office button—Save to save the table as HOURS WORKED.

Adding Records to a Table

At this point, you have set up the skeletons of three tables. The tables have no data records yet. If you printed the tables now, you would only see column headings (the field names). The most direct way to enter data into a table is to double-click the table's name in the navigation pane at the left side of the screen and type the data directly into the cells.

NOTE

To display and open the database objects, Access 2007 uses a navigation pane, which is on the left side of the Access window.

AT THE KEYBOARD

On the Database window's Home tab, double-click the EMPLOYEE table. Your data entry screen should resemble the one shown in Figure B-7.

Employee

Last Name	First Name	Employee ID	Street Address	City	State	Zip	Date Hired	US Citizen	Add New F
								☐	

FIGURE B-7 The data entry screen for the EMPLOYEE table

The EMPLOYEE table has many fields, some of which may be off the screen to the right. Scroll to see obscured fields. (Scrolling happens automatically as data is entered.) Figure B-7 shows all of the fields on the screen.

Type in your data one field value at a time. Note that the first row is empty when you begin. Each time you finish a value, press Enter; the cursor will move to the next cell. After data has been entered in the last cell in a row, the cursor moves to the first cell of the next row *and* Access automatically saves the record. (Thus, you do not need to perform the Office button—Save step after entering data into a table.)

When entering data in your table, note that dates (for example, in the Date Hired field) should be entered in the following format: 6/15/07. Access automatically expands the entry to the proper format in output.

Also note that Yes/No variables are clicked (checked) for Yes; otherwise, the box is left blank for No. You can change the box from Yes to No by clicking it, as if you were using a toggle switch.

Enter the data shown in Figure B-8 into the EMPLOYEE table. If you make errors in data entry, click the cell, backspace over the error, and type the correction.

Employee

Last Name	First Name	Employee ID	Street Address	City	State	Zip	Date Hired	US Citizen	Add New F
Howard	Jane	11411	28 Sally Dr	Glasgow	DE	19702	8/1/2009	☑	
Smith	John	12345	30 Elm St	Newark	DE	19711	6/1/1996	☑	
Smith	Albert	14890	44 Duce St	Odessa	DE	19722	7/15/1987	☑	
Jones	Sue	22282	18 Spruce St	Newark	DE	19716	7/15/2004	☐	
Ruth	Billy	71460	1 Tater Dr	Baltimore	MD	20111	8/15/1999	☐	
Add	Your	Data	Here	Elkton	MD	21921		☑	
								☐	

FIGURE B-8 Data for the EMPLOYEE table

Note that the sixth record is *your* data record. Assume that you live in Elkton, Maryland, were hired on today's date (enter the date), and are a U.S. citizen. Make up a fictitious Employee ID number. (For purposes of this tutorial, this sixth record has been created using the name of one of this text's authors and the employee ID 09911.)

After adding records to the EMPLOYEE table, open the WAGE DATA table and enter the data shown in Figure B-9.

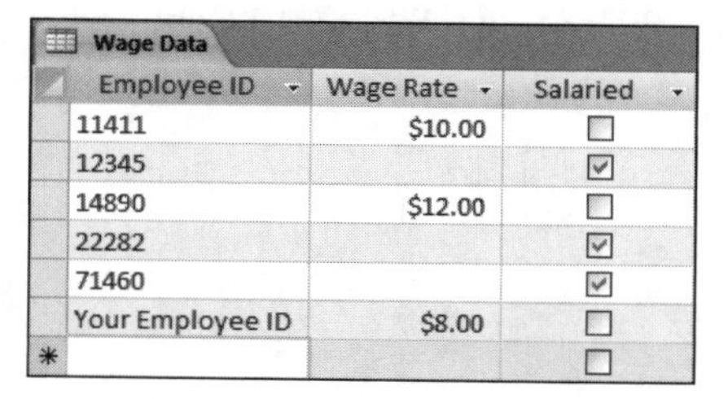

Wage Data

Employee ID	Wage Rate	Salaried
11411	$10.00	☐
12345		☑
14890	$12.00	☐
22282		☑
71460		☑
Your Employee ID	$8.00	☐
*		☐

FIGURE B-9 Data for the WAGE DATA table

In this table, you are again asked to create a new entry. For this record, enter your own employee ID. Also assume that you earn $8 an hour and are not salaried. (Note that when an employee's Salaried box is not checked (i.e., Salaried = No), the implication is that the employee is paid by the hour. Because salaried employees are not paid by the hour, their hourly rate is 0.00.)

When you finish creating the WAGE DATA table, open the HOURS WORKED table and enter the data shown in Figure B-10.

Hours Worked

Employee ID	Week #	Hours
11411	1	40
11411	2	50
12345	1	40
12345	2	40
14890	1	38
14890	2	40
22282	1	40
22282	2	40
71460	1	40
71460	2	40
Your Employee ID	1	60
Your Employee ID	2	55
*		

FIGURE B-10 Data for the HOURS WORKED table

Notice that salaried employees are always given 40 hours. Nonsalaried employees (including you) might work any number of hours. For your record, enter your fictitious employee ID, 60 hours worked for Week 1, and 55 hours worked for Week 2.

CREATING QUERIES

Because you know how to create basic queries, this section teaches you the kinds of advanced queries you will create in the Case Studies.

Using Calculated Fields in Queries

A **calculated field** is an output field made up of *other* field values. A calculated field is *not* a field in a table; it is created in the query generator. The calculated field does not become part of the table—it is just part of query output. The best way to understand this process is to work through an example.

AT THE KEYBOARD

Suppose you want to see the employee IDs and wage rates of hourly workers, and what the wage rates would be if all employees were given a 10 percent raise. To view that information, show the employee ID, the current wage rate, and the higher rate (which should be titled New Rate in the output). Figure B-11 shows how to set up the query.

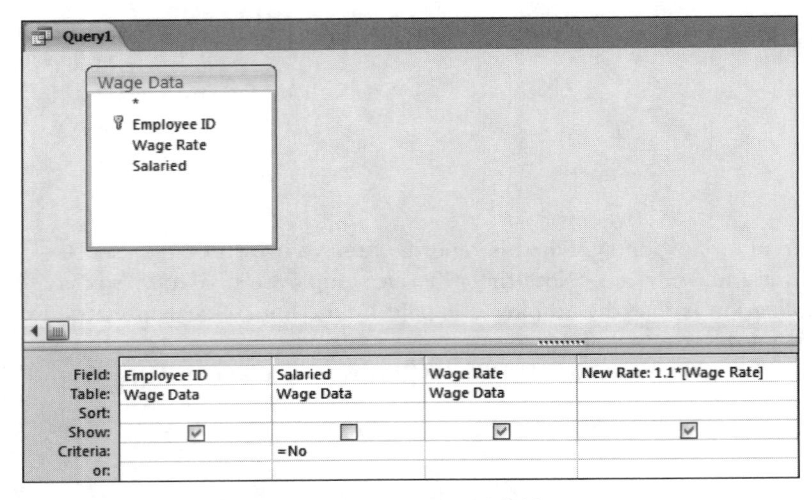

FIGURE B-11 Query setup for the calculated field

To set up this query, you need to select hourly workers by using the Salaried field with the Criteria = No. Note in Figure B-11 that the Show box for that field is not checked, so the Salaried field values will not appear in the query output.

Note the expression for the calculated field, which you can see in the rightmost field cell:

New Rate: 1.1 * [Wage Rate]

The term *New Rate:* merely specifies the desired output heading. (Don't forget the colon.) The rest of the expression, 1.1 * [Wage Rate], multiplies the old wage rate by 110 percent, which results in the 10 percent raise.

In the expression, the field name Wage Rate must be enclosed in square brackets. Remember this rule: *Any time an Access expression refers to a field name, the expression must be enclosed in square brackets.*

If you run this query, your output should resemble that in Figure B-12.

Query1

Employee ID	Wage Rate	New Rate
11411	$10.00	11
14890	$12.00	13.2
09911	$8.00	8.8

FIGURE B-12 Output for a query with calculated field

Notice that the calculated field output is not shown in Currency format, but as a Double—a number with digits after the decimal point. To convert the output to Currency format, you should select the output column by clicking the line above the calculated field expression, thus activating the column, which subsequently darkens. Your data entry screen should resemble the one shown in Figure B-13.

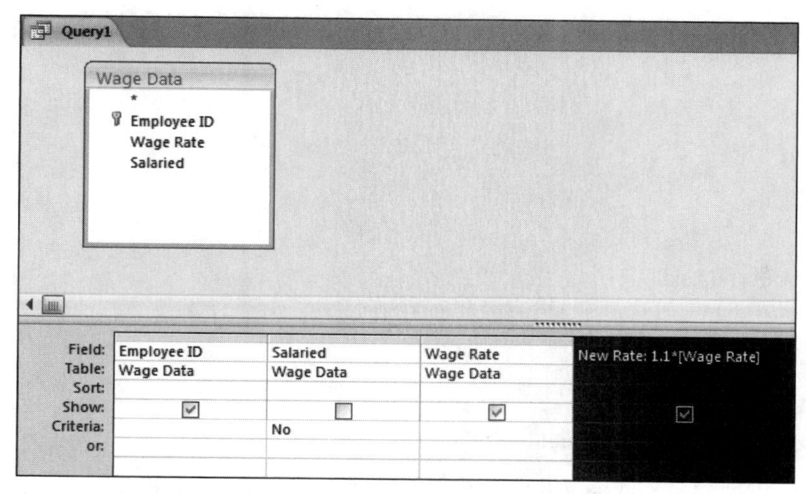

FIGURE B-13 Activating a calculated field in query design

Then, on the Design tab header, click Property Sheet in the Show/Hide group. A Field Properties window appears, as shown on the right in Figure B-14.

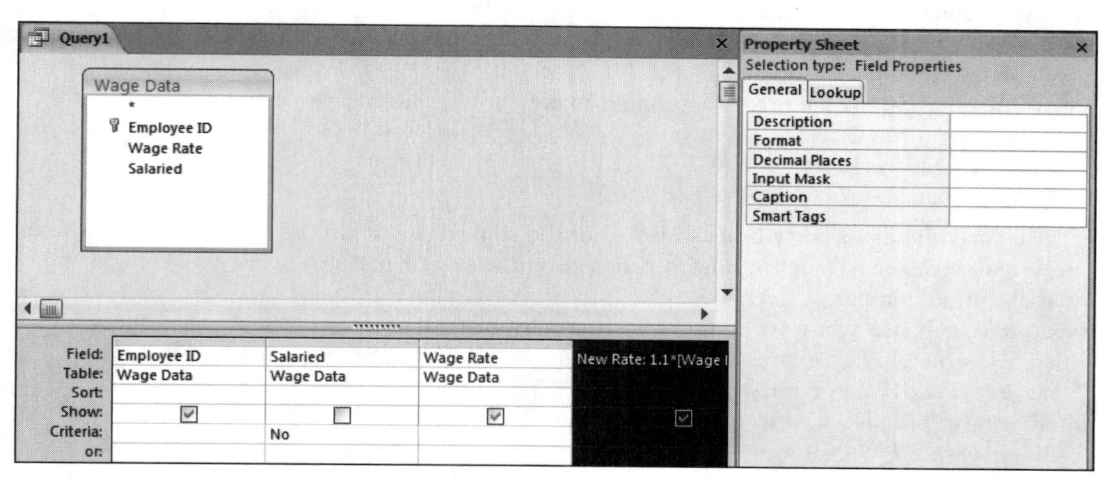

FIGURE B-14 Field Properties of a calculated field

Click Format and choose Currency, as shown in Figure B-15. Then click the X in the upper-right corner of the window to close it.

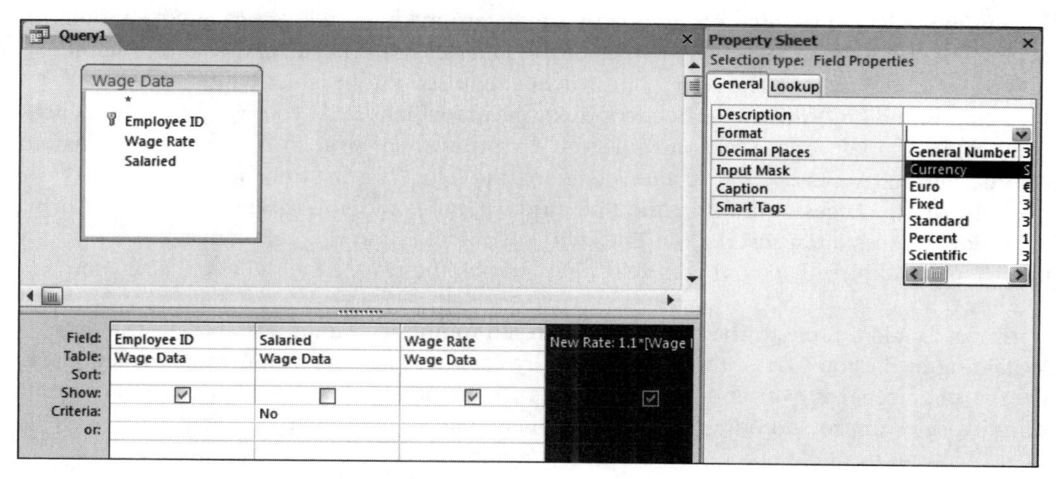

FIGURE B-15 Currency format of a calculated field

When you run the query, the output should resemble that in Figure B-16.

Employee ID	Wage Rate	New Rate
11411	$10.00	$11.00
14890	$12.00	$13.20
09911	$8.00	$8.80
*		

FIGURE B-16 Query output with formatted calculated field

Next, you'll examine how to avoid errors when making calculated fields.

Avoiding Errors When Making Calculated Fields

Follow these guidelines to avoid making errors in calculated fields:

- Don't put the expression in the *Criteria* cell as if the field definition were a filter. You are making a field, so put the expression in the *Field* cell.
- Spell, capitalize, and space a field's name *exactly* as you did in the table definition. If the table definition differs from what you type, Access thinks you're defining a new field by that name. Access then prompts you to enter values for the new field, which it calls a Parameter Query field. This problem is easy to debug because of the tag Parameter Query. If Access asks you to enter values for a parameter, you almost certainly misspelled a field name in an expression in a calculated field or criterion.
  For example, here are some errors you might make for Wage Rate:
  Misspelling: (Wag Rate)
  Case change: (wage Rate / WAGE RATE)
  Spacing change: (WageRate / Wage Rate)
- Don't use parentheses or curly braces instead of the square brackets. Also, don't put parentheses inside square brackets. You *can*, however, use parentheses outside the square brackets, in the normal algebraic manner.
  For example, suppose you want to multiply Hours by Wage Rate to get a field called Wages Owed. This is the correct expression:
  Wages Owed: [Wage Rate] * [Hours]
  The following expression also would be correct:
  Wages Owed: ([Wage Rate] * [Hours])
  But it would *not* be correct to omit the inside brackets, which is a common error:
  Wages Owed: [Wage Rate * Hours]

"Relating" Two (or More) Tables by the Join Operation

Often, the data you need for a query is in more than one table. To complete the query, you must **join** the tables by linking the common fields. One rule of thumb is that joins are made on fields that have common *values*, and those fields often can be key fields. The names of the join fields are irrelevant; also, the names of the tables (or fields) to be joined may be the same, but it is not required for an effective join.

Make a join by bringing in (adding) the tables needed. Next, decide which fields you will join. Then click one field name and hold down the left mouse button while you drag the cursor over to the other field's name in its window. Release the button. Access puts in a line, signifying the join. (If a relationship between two tables has been formed elsewhere, Access will insert the line automatically, and you will not have to perform the click-and-drag operation. Access often inserts join lines without the user forming relationships.)

You can join more than two tables. The common fields *need not* be the same in all tables; that is, you can daisy-chain them together.

A common join error is to add a table to the query and then fail to link it to another table. In that case, you will have a table floating in the top part of the QBE (query by example) screen. When you run the query, your output will show the same records over and over. That error is unmistakable because there is *so much* redundant output. The two rules are to add only the tables you need and to link all tables.

Next, you'll work through an example of a query that needs a join.

AT THE KEYBOARD

Suppose you want to see the last names, employee IDs, wage rates, salary status, and citizenship only for U.S. citizens and hourly workers. Because the data is spread across two tables, EMPLOYEE and WAGE DATA, you should add both tables and pull down the five fields you need. Then you should add the Criteria. Set up your work to resemble that in Figure B-17. Make sure the tables are joined on the common field, Employee ID.

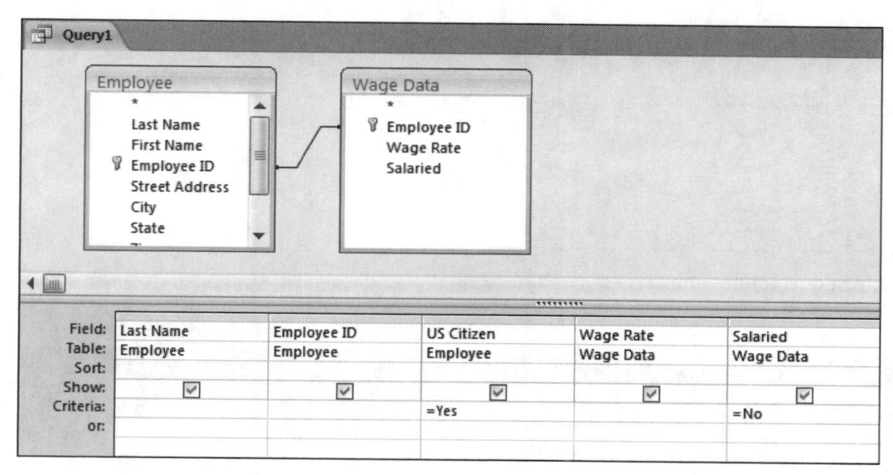

FIGURE B-17 A query based on two joined tables

You should quickly review the criteria you will need to set up this join: If you want data for employees who are U.S. citizens *and* who are hourly workers, the Criteria expressions go in the *same* Criteria row. If you want data for employees who are U.S. citizens *or* who are hourly workers, one of the expressions goes in the second Criteria row (the one that has the or: notation).

Now run the query. The output should resemble that in Figure B-18, with the exception of the name "Brady."

Query1

Last Name	Employee ID	US Citizen	Wage Rate	Salaried
Howard	11411	☑	$10.00	☐
Smith	14890	☑	$12.00	☐
Brady	09911	☑	$8.00	☐
*		☐		☐

FIGURE B-18 Output of a query based on two joined tables

You do not need to print the query output or save it, so return to Design view and close the query. Another practice query follows.

AT THE KEYBOARD

Suppose you want to see the wages owed to hourly employees for Week 2. You should show the last name, the employee ID, the salaried status, the week #, and the wages owed. Wages will have to be a calculated field ([Wage Rate] * [Hours]). The criteria are No for Salaried and 2 for the Week #. (This means another "And" query is required.) Your query should be set up like the one in Figure B-19.

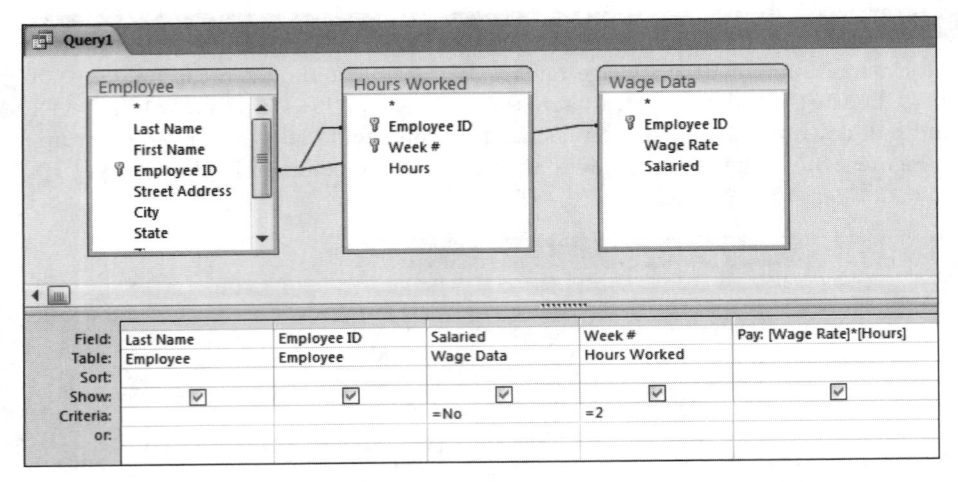

FIGURE B-19 Query setup for wages owed to hourly employees for Week 2

NOTE

In the query in Figure B-19, the calculated field column was widened so you can see the whole expression. To widen a column, click the column boundary line and drag to the right.

Run the query. The output should be similar to that in Figure B-20 (if you formatted your calculated field to Currency).

Query1

Last Name	Employee ID	Salaried	Week #	Pay
Howard	11411	☐	2	$500.00
Smith	14890	☐	2	$480.00
Brady	09911	☐	2	$440.00
*		☐		

FIGURE B-20 Query output for wages owed to hourly employees for Week 2

Notice that it was not necessary to pull down the Wage Rate and Hours fields to make the query work. You do not need to print the query output or save it, so return to Design view and close the query.

Summarizing Data from Multiple Records (Totals Queries)

You may want data that summarizes values from a field for several records (or possibly all records) in a table. For example, you might want to know the average hours that all employees worked in a week or perhaps the total (sum) of all the hours worked. Furthermore, you might want data grouped (stratified) in some way. For example, you might want to know the average hours worked, grouped by all U.S. citizens versus all non-U.S. citizens. Access calls such a query a **Totals query**. Those operations include the following:

Sum	The total of a given field's values
Count	A count of the number of instances in a field—that is, the number of records. (In the current example, you would count the number of employee IDs to get the number of employees.)
Average	The average of a given field's values
Min	The minimum of a given field's values
Var	The variance of a given field's values
StDev	The standard deviation of a given field's values
Where	The field has criteria for the query output

AT THE KEYBOARD

Suppose you want to know how many employees are represented in the example database. First, bring the EMPLOYEE table into the QBE screen. Because you will need to count the number of employee IDs, which is a Totals query operation, you must bring down the Employee ID field.

To tell Access you want a Totals query, click the Totals icon in the Design tab in the Show/Hide group. A new row called the Total row opens in the lower part of the QBE screen. At this point, the screen resembles that in Figure B-21.

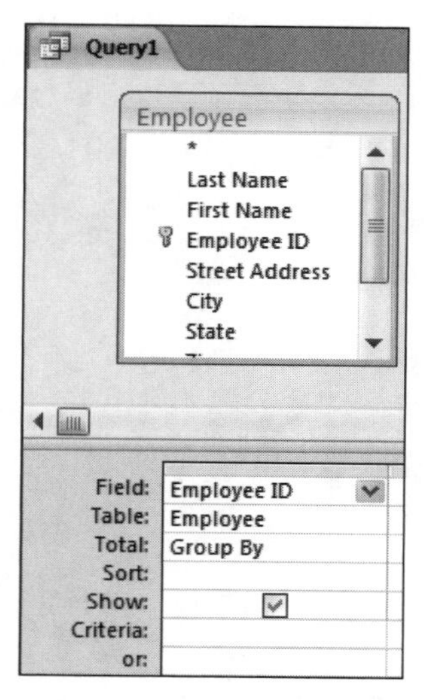

FIGURE B-21 Totals query setup

Note that the Total cell contains the words *Group By*. Until you specify a statistical operation, Access assumes that a field will be used for grouping (stratifying) data.

To count the number of employee IDs, click next to Group By, which reveals a little arrow. Click the arrow to reveal a drop-down menu, as shown in Figure B-22.

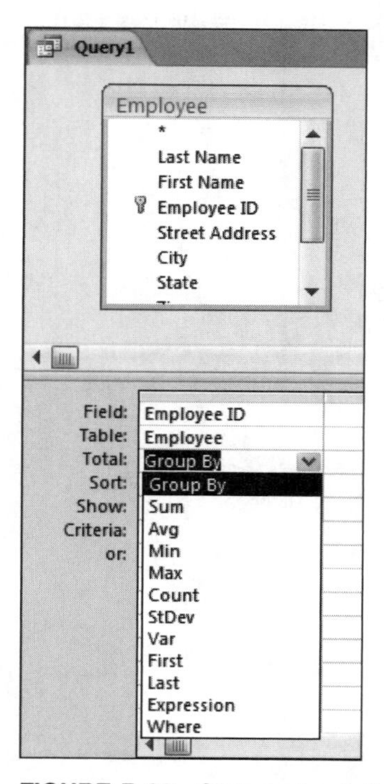

FIGURE B-22 Choices for statistical operation in a Totals query

Select the Count operator. (With this menu, you may need to scroll to see the operator you want.) Your screen should now resemble the one shown in Figure B-23.

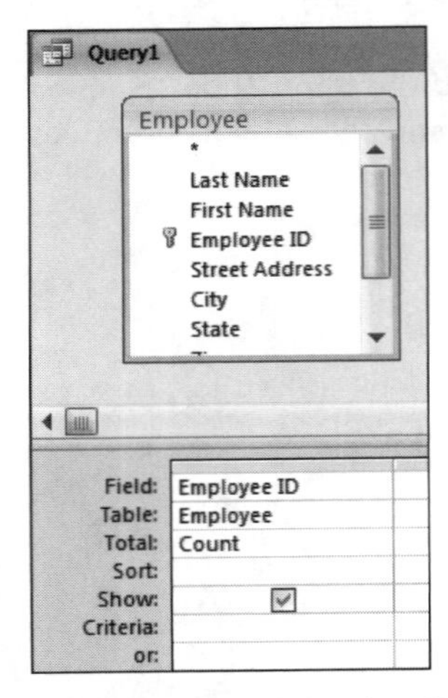

FIGURE B-23 Count in a Totals query

Run the query. Your output should resemble that in Figure B-24.

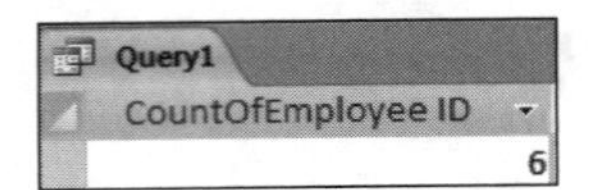

FIGURE B-24 Output of Count in a Totals query

Notice that Access made a pseudo-heading "CountOfEmployee ID" by splicing together the statistical operation (Count), the word *Of*, and the name of the field (Employee ID). What if you wanted a phrase such as *Count of Employees* as a heading? In Design view, you would change the query to resemble the one shown in Figure B-25.

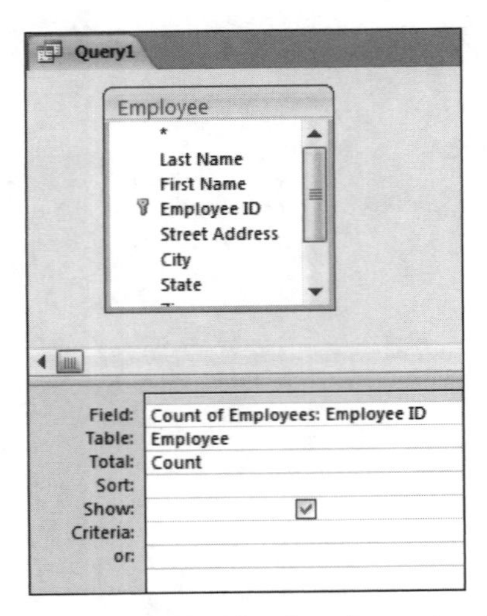

FIGURE B-25 Heading change in a Totals query

When you run the query, the output should resemble that in Figure B-26.

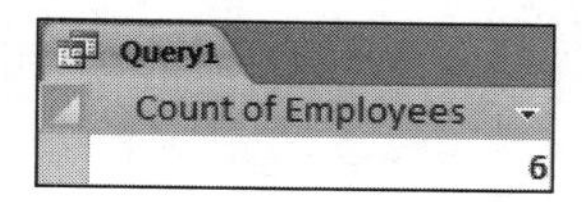

FIGURE B-26 Output of heading change in a Totals query

You do not need to print the query output or save it, so return to Design view and close the query.

AT THE KEYBOARD

As another example of a Totals query, suppose you want to know the average wage rate of employees, grouped by whether the employees are salaried. Figure B-27 shows how your query should be set up.

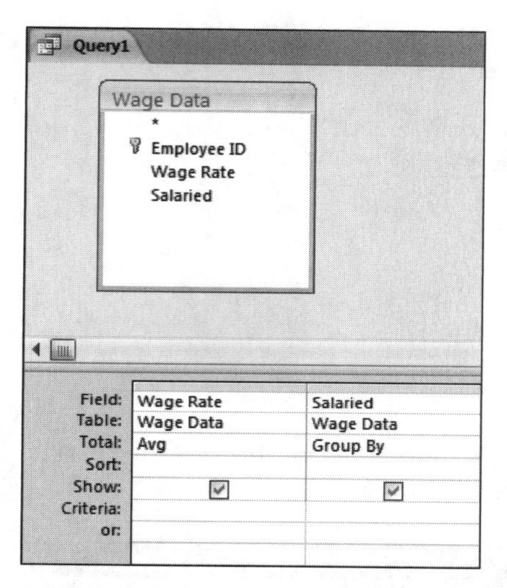

FIGURE B-27 Query setup for average wage rate of employees

When you run the query, your output should resemble that in Figure B-28.

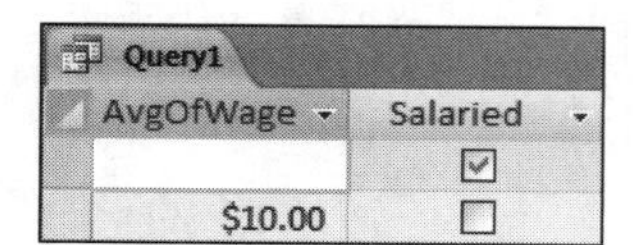

FIGURE B-28 Output of query for average wage rate of employees

Recall the convention that salaried workers are assigned zero dollars an hour. Suppose you want to eliminate the output line for zero dollars an hour because only hourly-rate workers matter for that query. The query setup is shown in Figure B-29.

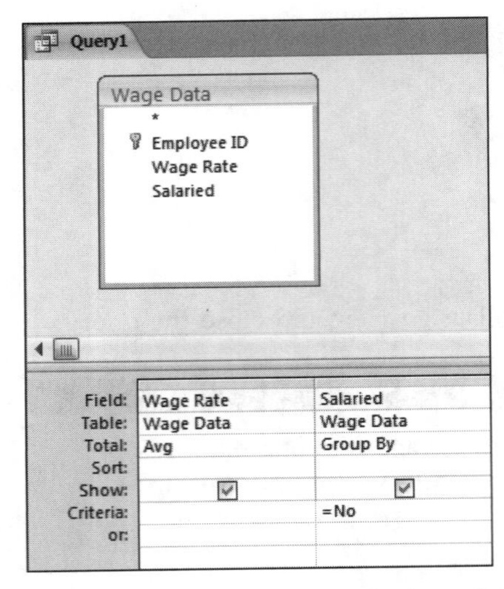

FIGURE B-29 Query setup for nonsalaried workers only

When you run the query, you'll get output for nonsalaried employees only, as shown in Figure B-30.

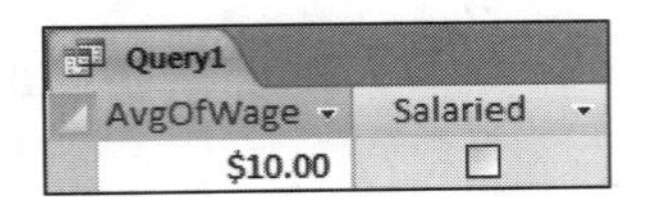

FIGURE B-30 Query output for nonsalaried workers only

Thus, it's possible to use Criteria in a Totals query, just as you would with a "regular" query. You do not need to print the query output or save it, so return to Design view and close the query.

AT THE KEYBOARD

Assume that you want to see two pieces of information for hourly workers: (1) the average wage rate—call it Average Rate in the output; and (2) 110 percent of the average rate—call it the Increased Rate. To get this information, you can make a calculated field in a Totals query.

You already know how to do certain things for this query. The revised heading for the average rate will be Average Rate (type *Average Rate: Wage Rate* in the Field cell). Note that you want the average of this field. Also, the grouping will be by the Salaried field. (To get hourly workers only, enter *Criteria: No*.)

The most difficult part of this query is to construct the expression for the calculated field. Conceptually, it is as follows:

Increased Rate: 1.1 * [The current average, however that is denoted]

The question is how to represent [The current average]. You cannot use Wage Rate because that heading denotes the wages before they are averaged. Surprisingly, you can use the new heading (Average Rate) to denote the averaged amount as follows:

Increased Rate: 1.1 * [Average Rate]

Although it may seem counterintuitive, *you can treat* Average Rate *as if it were an actual field name*. Note, however, that if you use a calculated field such as Average Rate in another calculated field, as shown in Figure B-31, you must show that original calculated field in the query output. If you don't, the query will ask you to *enter parameter value*, which is incorrect. Use the setup shown in Figure B-31.

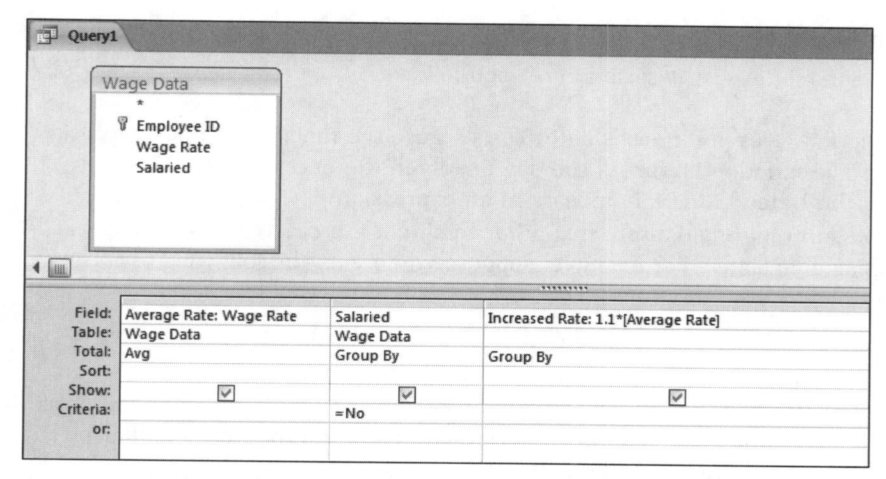

FIGURE B-31 Using a calculated field in another calculated field

NOTE

If you ran the query shown in Figure B-31, an error message would appear because no statistical operator is applied to the calculated field's Total cell; instead, the words Group By appear there. To correct that, you must change the Group By operator to Expression. You may have to scroll down to see Expression in the list.

Figure B-32 shows how the screen looks before the query is run.

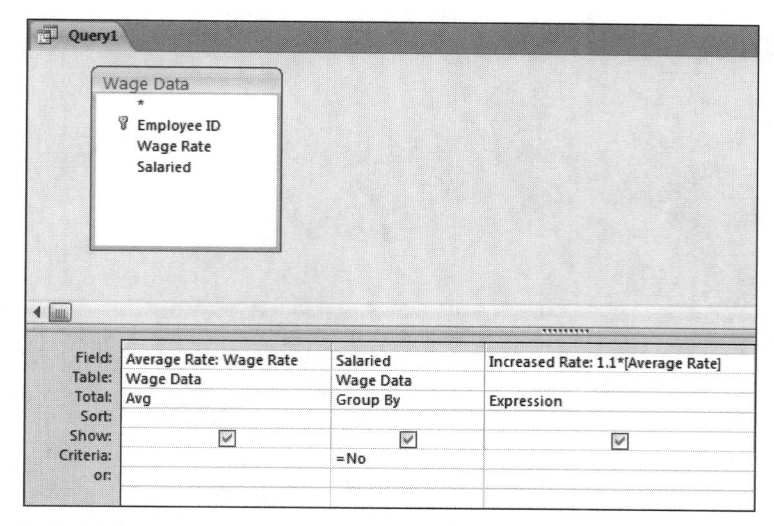

FIGURE B-32 An Expression in a Totals query

Figure B-33 shows the output of the query (note that the calculated field is formatted).

FIGURE B-33 Output of an Expression in a Totals query

You do not need to print the query output or save it, so return to Design view and close the query.

Using the Date( ) Function in Queries

Access has two important date function features. A description of each follows.

1. The built-in Date( ) function gives you today's date. You can use the function in query criteria or in a calculated field. The function "returns" the day on which the query is run; that is, it inserts the value where the Date( ) function appears in an expression.
2. *Date arithmetic* lets you subtract one date from another to obtain the difference—in number of days—between two calendar dates. For example, suppose you create the following expression: 10/9/2009 – 10/4/2009
   Access would evaluate the expression as the integer 5 (9 less 4 is 5).

As another example of how date arithmetic works, suppose you want to give each employee a bonus equaling a dollar for each day the employee has worked for you. You would need to calculate the number of days between the employee's date of hire and the day the query is run, and then multiply that number by $1.

You would find the number of elapsed days by using the following equation:

Date( ) – [Date Hired]

Also suppose that for each employee, you want to see the last name, employee ID, and bonus amount. You would set up the query as shown in Figure B-34.

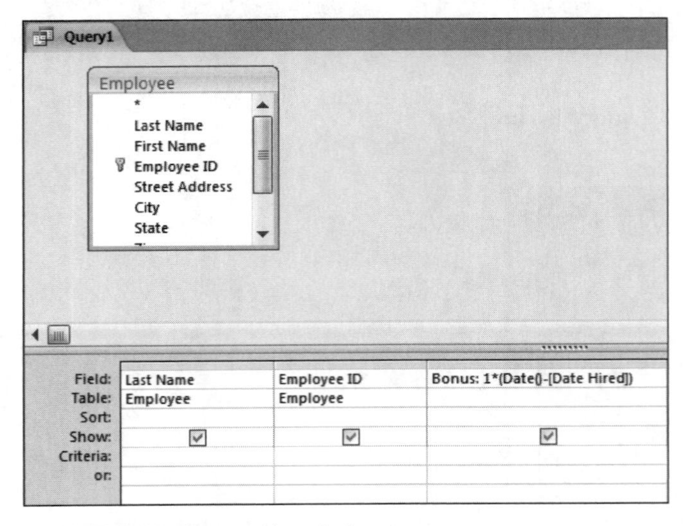

FIGURE B-34 Date arithmetic in a query

Assume you set the format of the Bonus field to Currency. The output will be similar to that in Figure B-35. (Your Bonus data will be different because the date differs from the date this tutorial was written.)

Last Name	Employee ID	Bonus
Brady	09911	$0.00
Howard	11411	$137.00
Smith	12345	$4,581.00
Smith	14890	$7,825.00
Jones	22282	$1,615.00
Ruth	71460	$3,411.00

FIGURE B-35 Output of query with date arithmetic

Using Time Arithmetic in Queries

Access also allows you to subtract the values of time fields to get an elapsed time. Assume your database has a JOB ASSIGNMENTS table showing the times that nonsalaried employees were at work during a day. The definition is shown in Figure B-36.

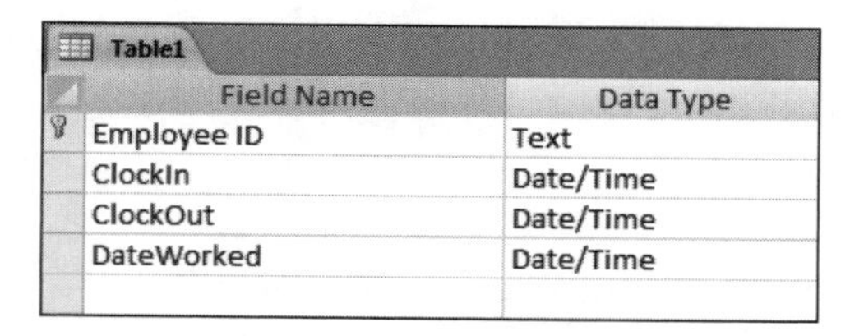

Table1

Field Name	Data Type
Employee ID	Text
ClockIn	Date/Time
ClockOut	Date/Time
DateWorked	Date/Time

FIGURE B-36 Date/Time data definition in the JOB ASSIGNMENTS table

Assume the DateWorked field is formatted for Long Date and the ClockIn and ClockOut fields are formatted for Medium Time. Also assume that for a particular day, nonsalaried workers were scheduled as shown in Figure B-37.

Job Assignments

Employee ID	ClockIn	ClockOut	DateWorked
09911	8:30:00 AM	4:30:00 PM	Thursday, September 30, 2010
11411	9:00:00 AM	3:00:00 PM	Thursday, September 30, 2010
14890	7:00:00 AM	5:00:00 PM	Thursday, September 30, 2010

FIGURE B-37 Display of date and time in a table

You want a query that shows the elapsed time that your employees were on the premises for the day. When you add the tables, your screen may show the links differently. Click and drag the JOB ASSIGNMENTS, EMPLOYEE, and WAGE DATA table icons to look like those in Figure B-38.

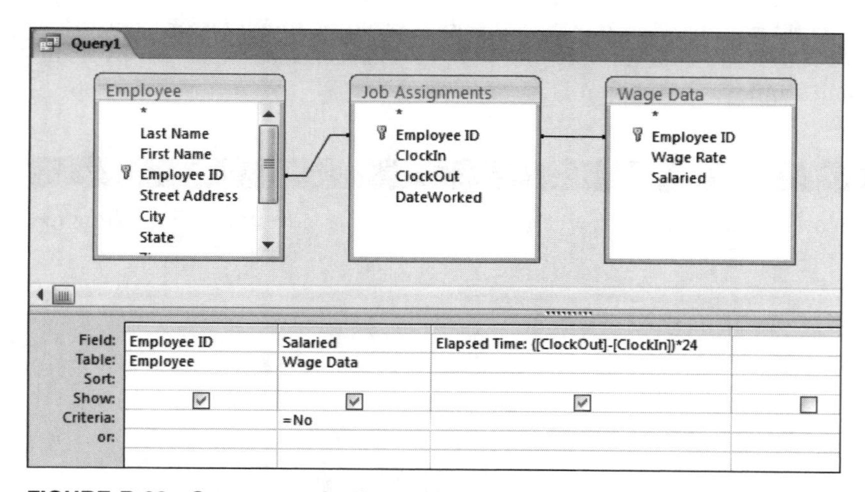

FIGURE B-38 Query setup for time arithmetic

Figure B-39 shows the output, which looks correct. For example, employee 09911 was at work from 8:30 a.m. to 4:30 p.m., which is eight hours. But how does the odd expression that follows yield the correct answers?

Query1

Employee ID	Salaried	Elapsed Time
09911	☐	8
11411	☐	6
14890	☐	10
*	☐	

FIGURE B-39 Query output for time arithmetic

([ClockOut] – [ClockIn]) * 24

Why wouldn't the following expression work?

[ClockOut] – [ClockIn]

Here is the answer: In Access, subtracting one time from the other yields the *decimal* portion of a 24-hour day. Returning to the example, you can see that employee 09911 worked eight hours, which is one-third of a day, so the time arithmetic function yields .3333. That is why you must multiply by 24—to convert from decimals to an hourly basis. Hence, for employee 09911, the expression performs the following calculation: $1/3 \times 24 = 8$.

Note that parentheses are needed to force Access to do the subtraction *first*, before the multiplication. Without parentheses, multiplication takes precedence over subtraction. For example, consider the following expression:

[ClockOut] – [ClockIn] * 24

In that case, ClockIn would be multiplied by 24, the resulting value would be subtracted from ClockOut, and the output would be a nonsensical decimal number.

Deleting and Updating Queries

The queries presented in this tutorial thus far have been Select queries. They select certain data from specific tables based on a given criterion. You also can create queries to update the original data in a database. Businesses do that often, and they do it in real time. For example, when you order an item from a Web site, the company's database is updated to reflect the purchase of the item through the deletion of that item from the company's inventory.

Consider an example. Suppose you want to give all of the nonsalaried workers a $0.50 per hour pay raise. With the three nonsalaried workers you have now, it would be easy to go into the table and simply change the Wage Rate data. But assume you have 3,000 nonsalaried employees. Now it would be much faster—not to mention more accurate—to change the Wage Rate data for each of the 3,000 nonsalaried employees by using an Update query that adds $0.50 to each employee's wage rate.

AT THE KEYBOARD

Now you will change each of the nonsalaried employees' pay via an Update query. Figure B-40 shows how to set up the query.

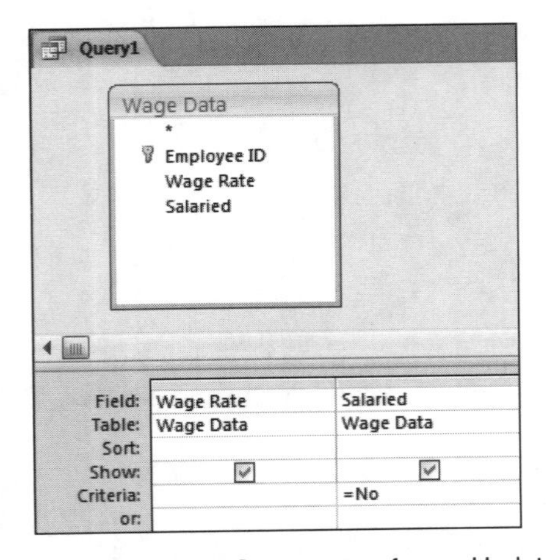

FIGURE B-40 Query setup for an Update query

So far, this query is just a Select query. Click the Update button in the Query Type group, as shown in Figure B-41.

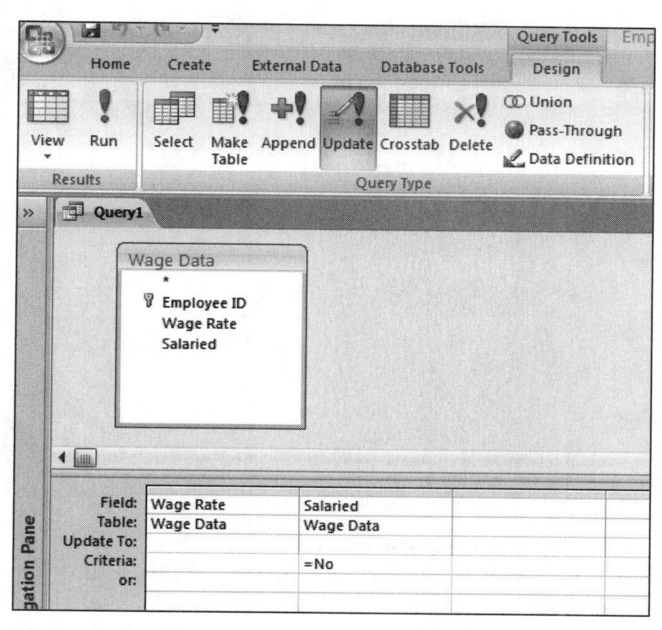

FIGURE B-41 Selecting a query type

Notice that you now have another line on the QBE grid called *Update To:*, which is where you specify the change or update to the data. Notice that you will update only the nonsalaried workers by using a filter under the Salaried field. Update the Wage Rate data to Wage Rate plus $0.50, as shown in Figure B-42. (Note that the update involves the use of brackets [ ], as in a calculated field.)

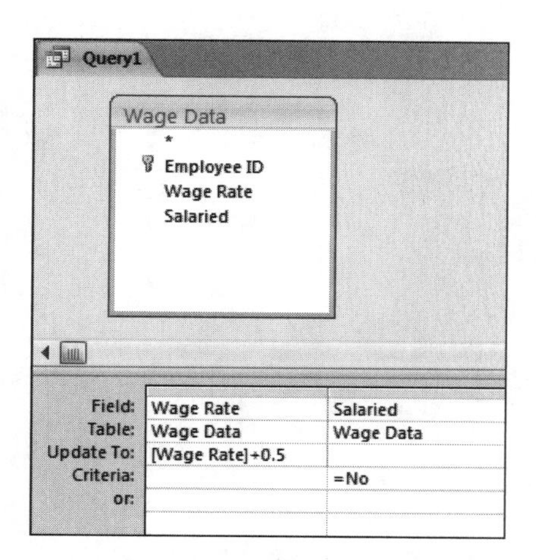

FIGURE B-42 Updating the wage rate for nonsalaried workers

Now run the query by clicking the Run button in the Results group. If you cannot run the query because it is blocked by Disabled Mode, choose Database Tools tab, Message Bar in the Show/Hide group. Click the Options button and choose "enable this content," then click OK. When you successfully run the query, the warning message shown in Figure B-43 appears.

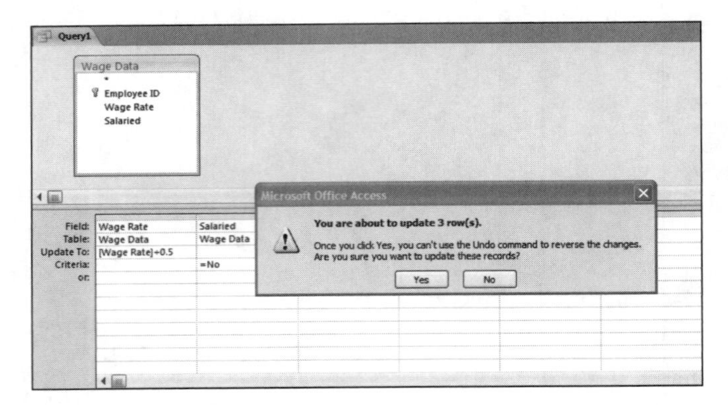

FIGURE B-43 Update query warning

When you click Yes, the records are updated. Check the updated records by viewing the WAGE DATA table. Each nonsalaried wage rate should be increased by $0.50. You could add or subtract data from another table as well. If you do, remember to put the field name in square brackets.

Another kind of query is the Delete query. Delete queries work like Update queries. For example, assume your company has been taken over by the state of Delaware. The state has a policy of employing only Delaware residents. Thus, you must delete (or fire) all employees who are not exclusively Delaware residents. To do that, you need to create a Select query. Using the EMPLOYEE table, you would click the Delete icon from the Query Type group, then bring down the State field and filter only those records not in Delaware (DE). Do not perform the operation, but note that if you did, the setup would look like the one in Figure B-44.

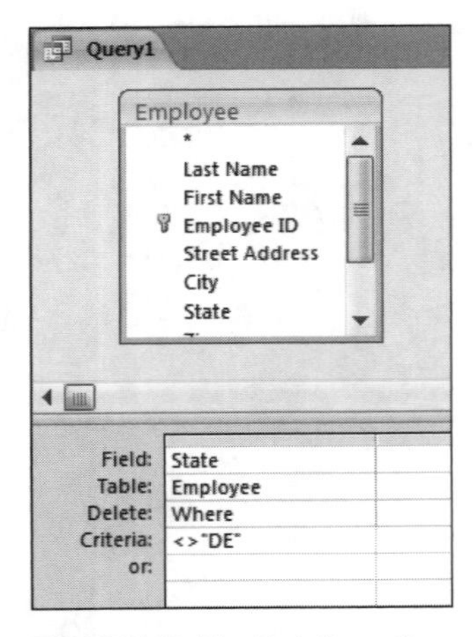

FIGURE B-44 Deleting all employees who are not Delaware residents

Using Parameter Queries

Another kind of query, which is actually a type of Select query, is a **Parameter query**. For example, suppose your company has 5,000 employees and you want to query the database to find the same kind of information again and again, but about different employees each time. For example, you might want to query the database to find out how many hours a particular employee has worked. To do that, you could run a query that you created and stored previously, but run it only for a particular employee.

AT THE KEYBOARD

Create a Select query with the format shown in Figure B-45.

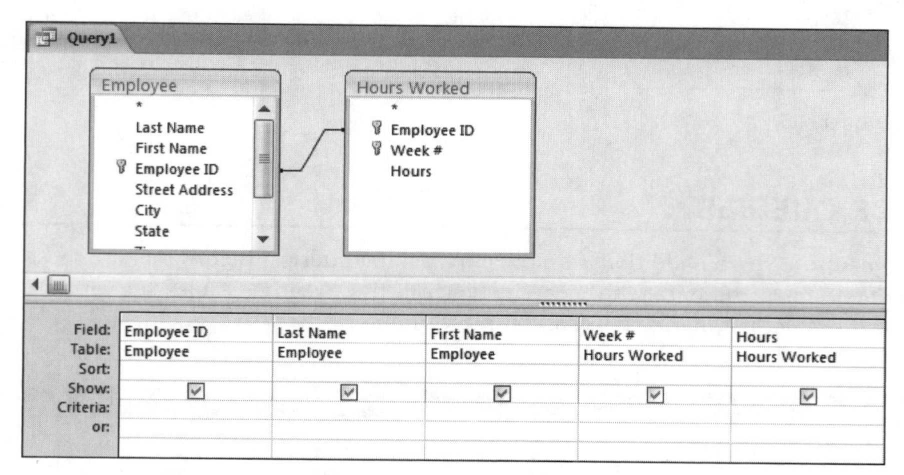

FIGURE B-45 Design of a Parameter query begins as a Select query

In the Criteria line of the QBE grid for the Employee ID field, type what is shown in Figure B-46.

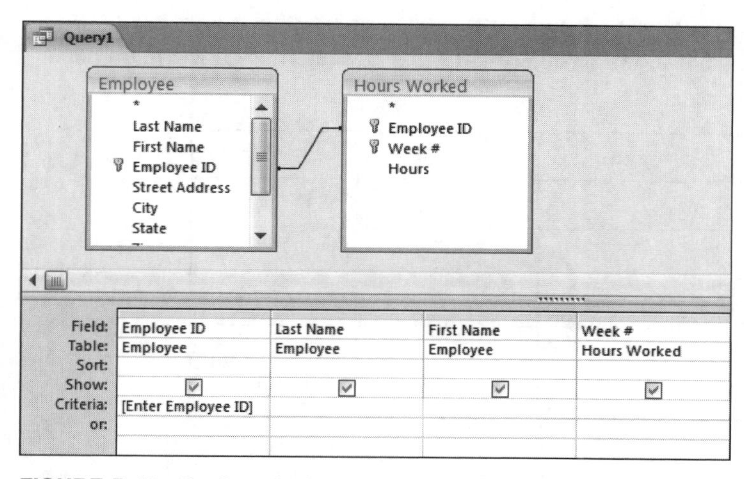

FIGURE B-46 Design of a Parameter query, continued

Note that the Criteria line involves the use of square brackets, as you would expect to see in a calculated field.

Now run the query. You will be prompted for the specific employee's ID number, as shown in Figure B-47.

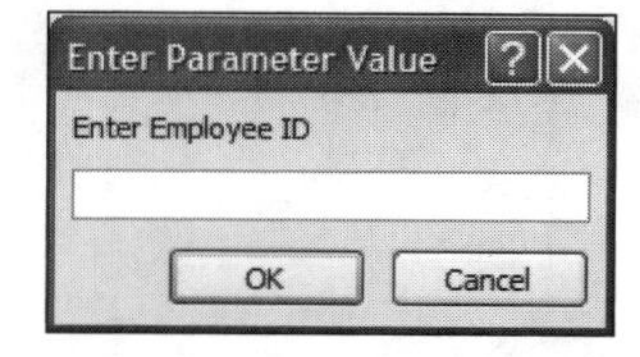

FIGURE B-47 Enter Parameter Value dialog box

Enter your own employee ID. Your query output should resemble that in Figure B-48.

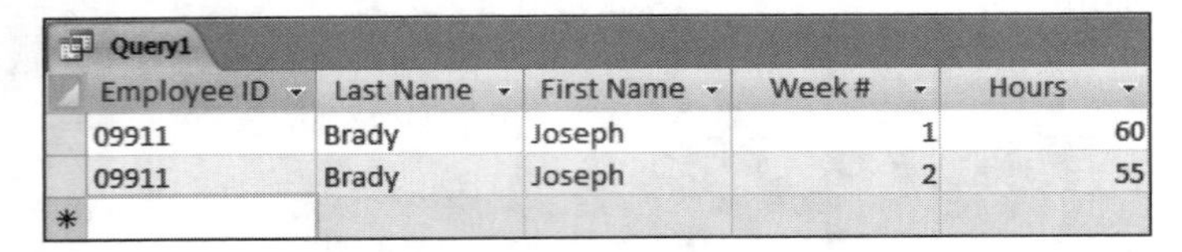

Query1

Employee ID	Last Name	First Name	Week #	Hours
09911	Brady	Joseph	1	60
09911	Brady	Joseph	2	55
*				

FIGURE B-48 Output of a Parameter query

MAKING SEVEN PRACTICE QUERIES

This portion of the tutorial is designed to give you additional practice in making queries. Before making these queries, you must create the specified tables and enter the records shown in the Creating Tables section of this tutorial. The output shown for the practice queries is based on those inputs.

AT THE KEYBOARD

For each query that follows, you are given a problem statement and a "scratch area." You also are shown what the query output should look like. Follow this procedure: Set up a query in Access. Run the query. When you are satisfied with the results, save the query and continue with the next query. Note that you will work with the EMPLOYEE, HOURS WORKED, and WAGE DATA tables.

1. Create a query that shows the employee ID, last name, state, and date hired for employees who live in Delaware *and* were hired after 12/31/99. Perform an ascending sort by employee ID. (To recap sorting procedures, first click the Sort cell of the field, and then choose Ascending or Descending.) Before creating your query, use the table shown in Figure B-49 to work out your QBE grid on paper.

Field					
Table					
Sort					
Show					
Criteria					
Or:					

FIGURE B-49 QBE grid template

Your output should resemble that in Figure B-50.

Query1

Employee ID	Last Name	State	Date Hired
11411	Howard	DE	8/1/2009
22282	Jones	DE	7/15/2004
*			

FIGURE B-50 Number 1 query output

2. Create a query that shows the last name, first name, date hired, and state for employees who live in Delaware *or* were hired after 12/31/99. The primary sort (ascending) is on last name, and the secondary sort (ascending) is on first name. (To recap, the Primary Sort field must be to the left of the Secondary Sort field in the query setup.) Before creating your query, use the table shown in Figure B-51 to work out your QBE grid on paper.

Field					
Table					
Sort					
Show					
Criteria					
Or:					

FIGURE B-51 QBE grid template

If your name was Joe Brady, your output would look like that in Figure B-52.

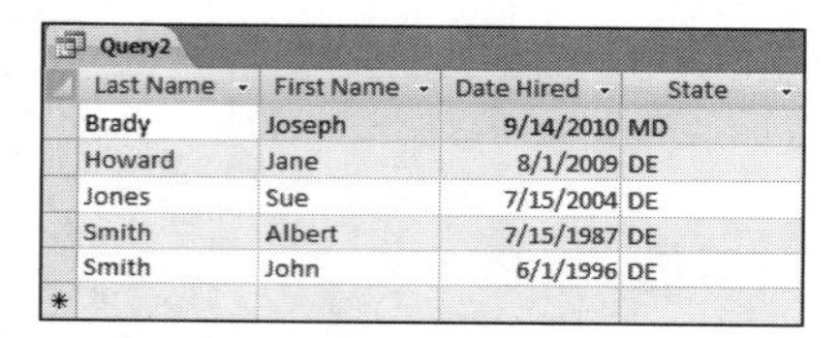
Query2

Last Name	First Name	Date Hired	State
Brady	Joseph	9/14/2010	MD
Howard	Jane	8/1/2009	DE
Jones	Sue	7/15/2004	DE
Smith	Albert	7/15/1987	DE
Smith	John	6/1/1996	DE

FIGURE B-52 Number 2 query output

3. Create a query that shows the sum of hours that U.S. citizens worked and the same sum for non-U.S. citizens (that is, group on citizenship). The heading for total hours worked should be Total Hours Worked. Before creating your query, use the table shown in Figure B-53 to work out your QBE grid on paper.

Field					
Table					
Total					
Sort					
Show					
Criteria					
Or:					

FIGURE B-53 QBE grid template

Your output should resemble that in Figure B-54.

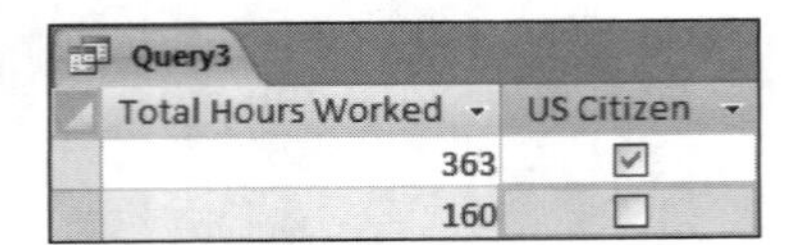

Query3

Total Hours Worked	US Citizen
363	☑
160	☐

FIGURE B-54 Number 3 query output

4. Create a query that shows the wages owed to hourly workers for Week 1. The heading for the wages owed should be Total Owed. The output headings should be Last Name, Employee ID, Week #, and Total Owed. Before creating your query, use the table shown in Figure B-55 to work out your QBE grid on paper.

Field					
Table					
Sort					
Show					
Criteria					
Or:					

FIGURE B-55 QBE grid template

If your name was Joe Brady, your output would look like that in Figure B-56.

Query4

Last Name	Employee ID	Week #	Total Owed
Howard	11411	1	$420.00
Smith	14890	1	$475.00
Brady	09911	1	$510.00
*			

FIGURE B-56 Number 4 query output

5. Create a query that shows the last name, employee ID, hours worked, and overtime amount owed for hourly employees who earned overtime during Week 2. Overtime is paid at 1.5 times the normal hourly rate for all the hours worked over 40. Note that the amount shown in the query should be just the overtime portion of the wages paid. Also, this is not a Totals query—amounts should be shown for individual workers. Before creating your query, use the table shown in Figure B-57 to work out your QBE grid on paper.

Field					
Table					
Sort					
Show					
Criteria					
Or:					

FIGURE B-57 QBE grid template

If your name was Joe Brady, your output would look like that in Figure B-58.

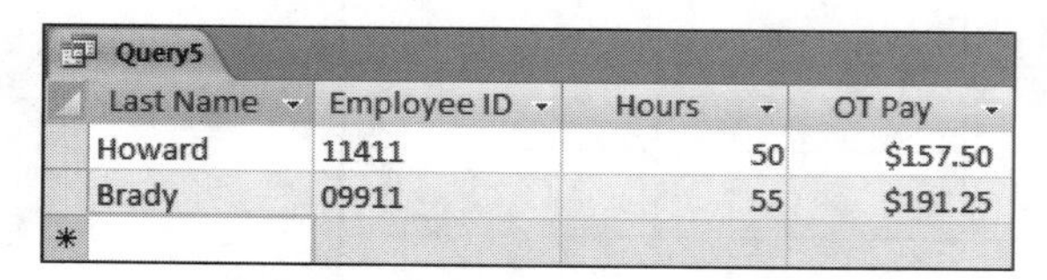
Query5

Last Name	Employee ID	Hours	OT Pay
Howard	11411	50	$157.50
Brady	09911	55	$191.25
*			

FIGURE B-58 Number 5 query output

6. Create a Parameter query that shows the hours employees have worked. Have the Parameter query prompt for the week number. The output headings should be Last Name, First Name, Week #, and Hours. Do this only for the nonsalaried workers. Before creating your query, use the table shown in Figure B-59 to work out your QBE grid on paper.

Field					
Table					
Sort					
Show					
Criteria					
Or:					

FIGURE B-59 QBE grid template

Run the query using 2 when prompted for the Week #. Your output should look like that in Figure B-60.

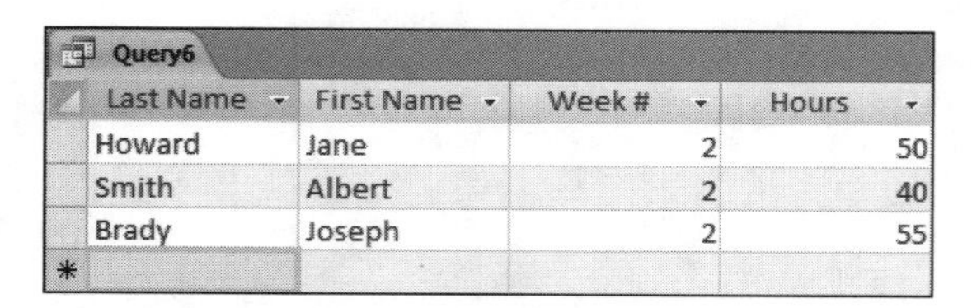
Query6

Last Name	First Name	Week #	Hours
Howard	Jane	2	50
Smith	Albert	2	40
Brady	Joseph	2	55
*			

FIGURE B-60 Number 6 query output

7. Create an update query that gives certain workers a merit raise. First, you must create an additional table as shown in Figure B-61.

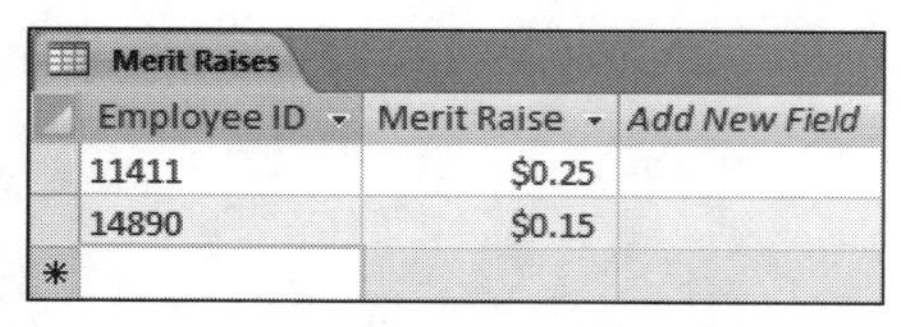
Merit Raises

Employee ID	Merit Raise	Add New Field
11411	$0.25	
14890	$0.15	
*		

FIGURE B-61 MERIT RAISES table

8. Make a query that adds the Merit Raise to the current Wage Rate for employees who will receive a raise. When you run the query, you should be prompted with *You are about to update two rows.* Check the original WAGE DATA table to confirm the update. Before creating your query, use the table shown in Figure B-62 to work out your QBE grid on paper.

Field					
Table					
Update to					
Criteria					
Or:					

FIGURE B-62 QBE grid template

CREATING REPORTS

Database packages let you make attractive management reports from a table's records or from a query's output. If you are making a report from a table, the Access report generator looks up the data in the table and puts it into report format. If you are making a report from a query's output, Access runs the query in the background (you do not control it or see it happen) and then puts the output in report format.

There are different ways to make a report. One method is to hand-craft the report from scratch in Design view, but this tedious process is not shown in this tutorial. A simpler way is to select the query or table on which the report is based and then click Create Report. This streamlined method of creating reports is new in the Access 2007 version and is shown in this tutorial.

Creating a Grouped Report

This tutorial assumes that you already know how to make a basic ungrouped report, so this section teaches you how to make a grouped report. (If you don't know how to make an ungrouped report, you can learn by following the first example in the upcoming section.)

AT THE KEYBOARD

Suppose you want to make a report out of the HOURS WORKED table. First select the table (do not open it, just click once to select it). Choose the Create tab, Report in the Reports group. A report appears, as shown in Figure B-63.

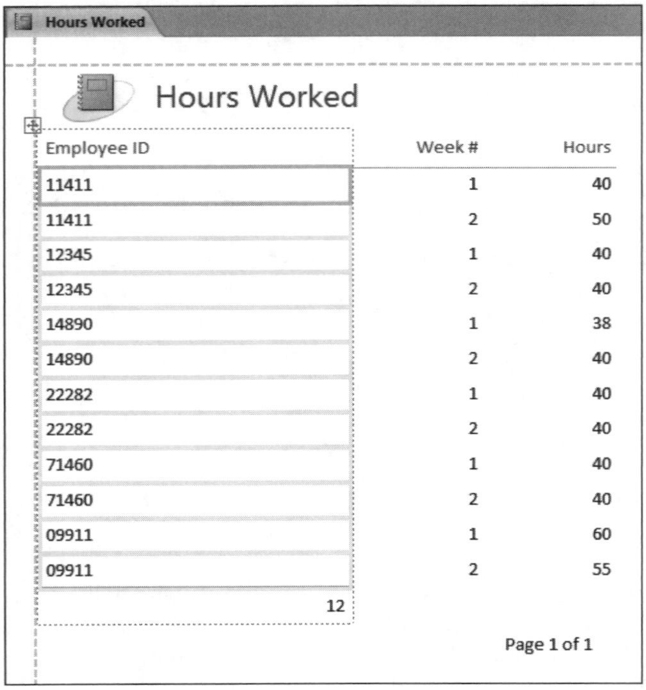

Hours Worked

Hours Worked

Employee ID	Week #	Hours
11411	1	40
11411	2	50
12345	1	40
12345	2	40
14890	1	38
14890	2	40
22282	1	40
22282	2	40
71460	1	40
71460	2	40
09911	1	60
09911	2	55
12		

Page 1 of 1

FIGURE B-63 Initial report based on a table

In the Format tab, Grouping and Totals Group, select the Group and Sort button. Your report will have an additional selection at the bottom, as shown in Figure B-64.

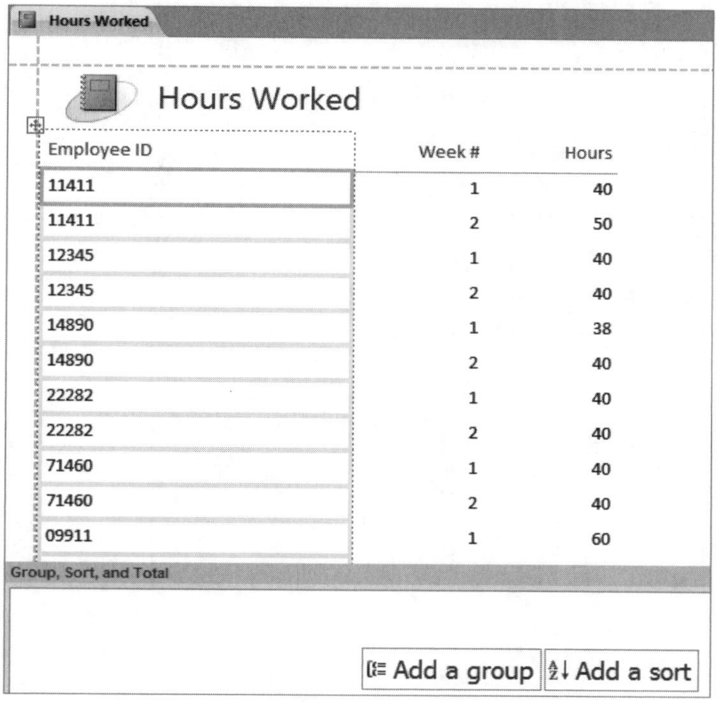

Hours Worked

Hours Worked

Employee ID	Week #	Hours
11411	1	40
11411	2	50
12345	1	40
12345	2	40
14890	1	38
14890	2	40
22282	1	40
22282	2	40
71460	1	40
71460	2	40
09911	1	60

Group, Sort, and Total

Add a group Add a sort

FIGURE B-64 Report with Grouping and Sorting options

Click the Add a group button at the bottom of the report, and select Employee ID. Your report will be grouped as shown in Figure B-65.

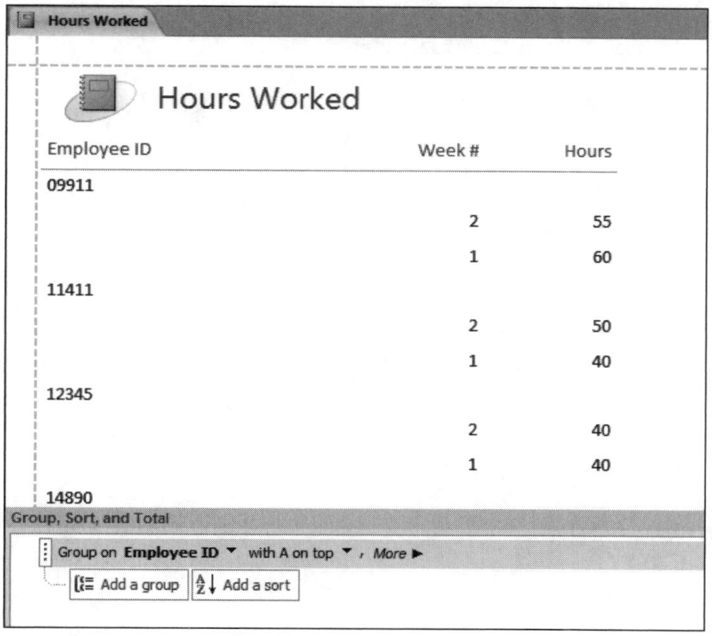

Hours Worked

Hours Worked

Employee ID	Week #	Hours
09911		
	2	55
	1	60
11411		
	2	50
	1	40
12345		
	2	40
	1	40
14890		

Group, Sort, and Total

Group on **Employee ID** with A on top , *More*

Add a group Add a sort

FIGURE B-65 Grouped report

To complete this report, you need to total the hours for each individual employee by selecting the Hours column heading. Your report will show that the entire column is selected. On the Format tab, Grouping and Totals Group, click the Totals button and then choose Sum from the drop-down menu, as shown in Figure B-66.

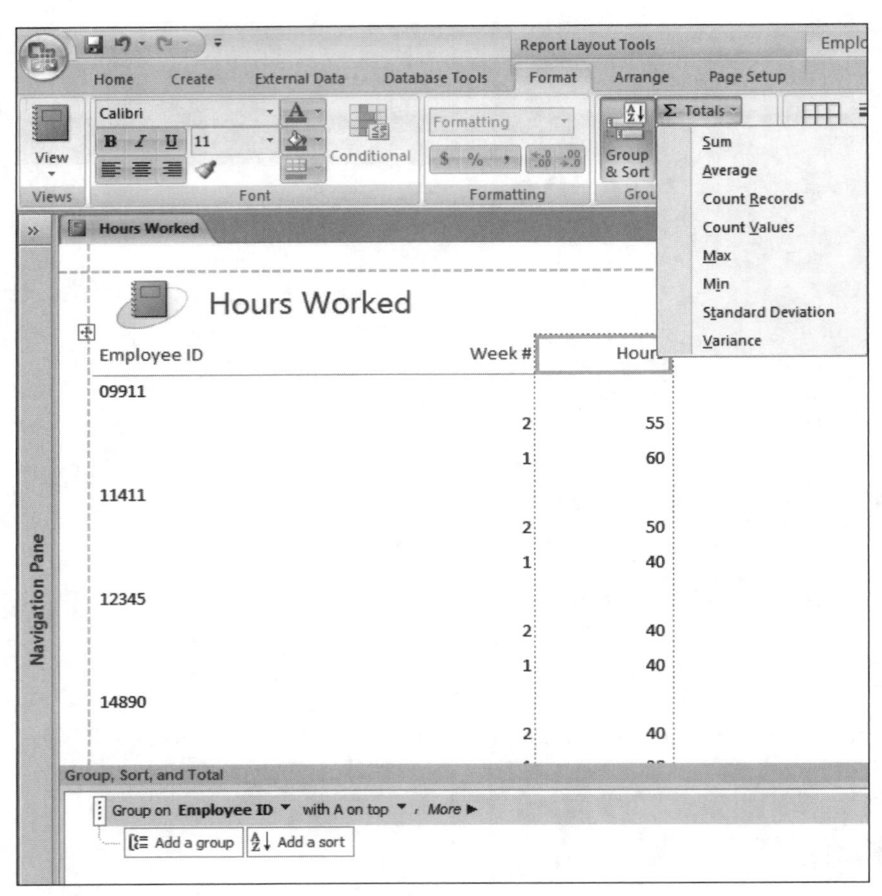

FIGURE B-66 Hours column selected

Your report will look like the one in Figure B-67.

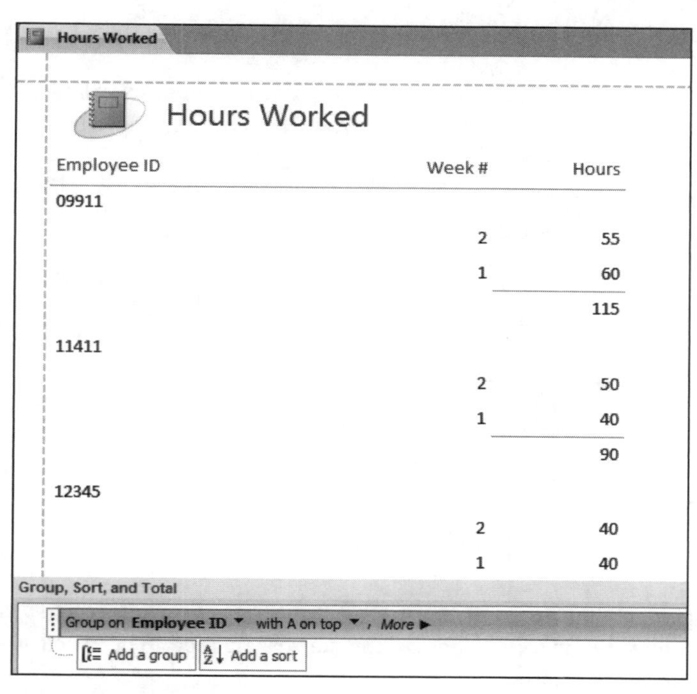

FIGURE B-67 Completed report

You are currently in Layout view of the report. To see how the final report looks when printed, choose the Format tab, Views Group, Report view. Your report looks like the one in Figure B-68 (only a portion appears).

Hours Worked

Hours Worked

Employee ID	Week #	Hours
09911		
	2	55
	1	60
		115
11411		
	2	50
	1	40
		90
12345		
	2	40
	1	40
		80

FIGURE B-68 Report in Report view

NOTE

It's easy to change the picture or logo at the upper-left corner of the report. Simply click the notebook symbol and press the Delete key. You can insert a logo in place of the notebook by choosing Format tab, Controls Group, Logo.

Moving Fields in Design View

If you group on more than one field in a report, the report will have an odd "staircase" look or display repeated data, or it will have both problems. Next, you will learn how to overcome these problems in Design view.

Suppose you make a query that shows an employee's last name, first name, week #, and hours worked, and then you make a report from that query, grouping on last name only. See Figure B-69.

Hours Worked by Employees

Wednesday, April 28, 2010
2:43:21 PM

Last Name	First Name	Week #	Hours
Brady			
	Joseph	2	55
	Joseph	1	60
Howard			
	Jane	2	50
	Jane	1	40
Jones			
	Sue	2	40
	Sue	1	40

FIGURE B-69 Query-based report grouped on last name

As you preview the report, notice the repeating data from the First Name field. In the report displayed in Figure B-69, notice that the first name repeats for each week worked—hence, the staircase effect. (The Week # and Hours fields are shown as subordinate to Last Name, as desired.)

Suppose you want the last name and first name to appear on the same line. If so, take the report into Design view for editing. To do this from the Format tab, choose Views Group, Design view. At this point, your report should look like the one in Figure B-70.

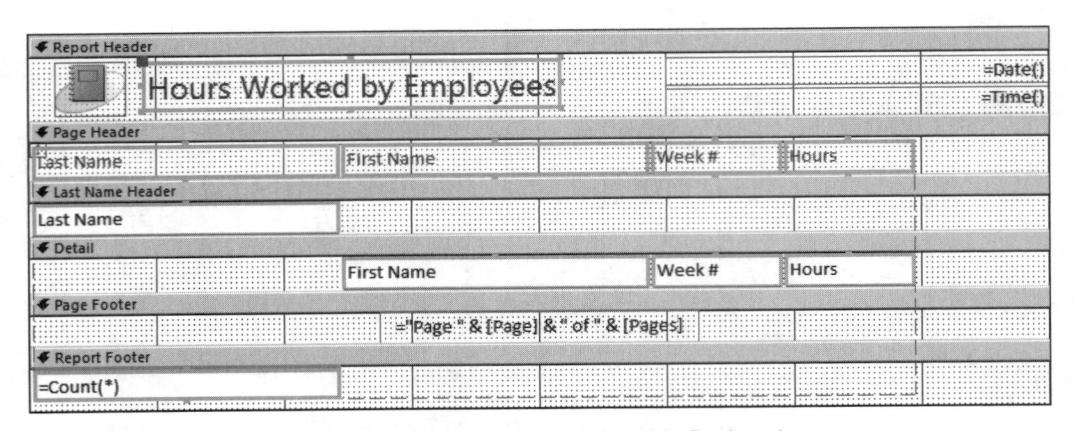

FIGURE B-70 Hours Worked by Employees report displayed in Design view

This complex screen has a hierarchical organization. At the top is the Report level, and the next level down (within a report) is the Page level. Below the Page level are the data grouping and the details of each group. The Report header is usually the title you specified. The Page header is usually the names of the fields you told Access to put in the report. The page number is inserted by default.

Your goal is to get the First Name into the Last Name header band (*not* into the Page Header band) so they will print on the same line. The first step is to click the First Name object in the Detail band, as shown in Figure B-71.

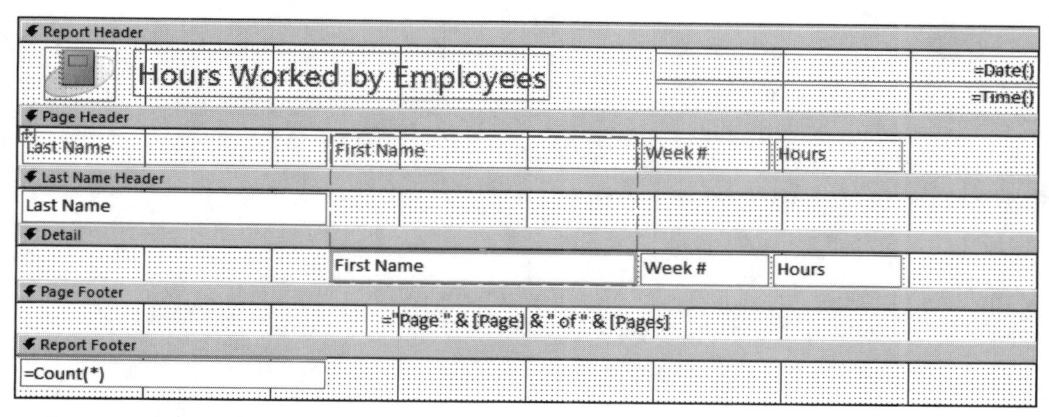

FIGURE B-71 Selecting First Name object in the Detail band

Right-click the First Name object, choose Layout, and then choose Move Up a Section, as shown in Figure B-72.

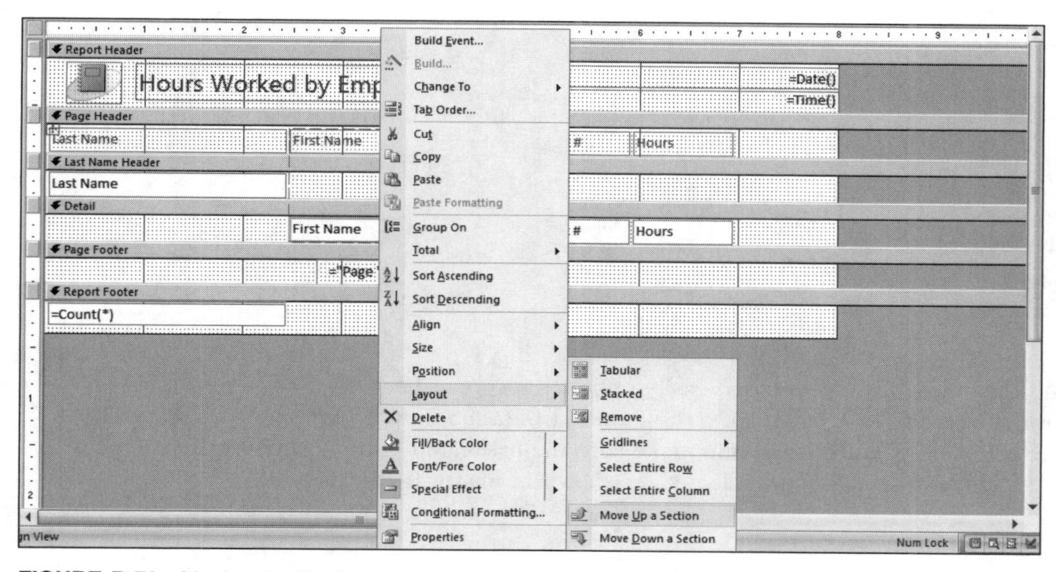

FIGURE B-72 Moving the First Name object to the Last Name header

View the report by choosing Views Group, Report view. The report shows the First Name on the same line as Last Name (see Figure B-73).

Hours Worked by Employees

Wednesday, April 28, 2010
2:44:06 PM

Last Name	First Name	Week #	Hours
Brady	Joseph		
		2	55
		1	60
Howard	Jane		
		2	50
		1	40
Jones	Sue		
		2	40
		1	40

FIGURE B-73 Report with Last Name and First Name on same line

You can now add the sum of Hours for each group. Also, if you want to add more fields to your report, such as Street Address and Zip, you can repeat the preceding procedure.

IMPORTING DATA

Text or spreadsheet data is easy to import into Access. In business, it is often necessary to import data because companies use disparate systems. For example, assume that your healthcare coverage data is on the human resources manager's computer in a Microsoft Excel spreadsheet. Open the Excel application and then create a spreadsheet using the data shown in Figure B-74.

	A	B	C
1	Employee ID	Provider	Level
2	11411	BlueCross	family
3	12345	BlueCross	family
4	14890	Coventry	spouse
5	22282	None	none
6	71460	Coventry	single
7	09911	BlueCross	single

FIGURE B-74 Excel data

Save the file and then close it. Now you can easily import the spreadsheet data into a new table in Access. With your Employee database open, go to the External Data tab, Import group and click Excel. Browse to find the Excel file you just created and make sure the first radio button ("Import the source data into a new table in the current database") is selected, as shown in Figure B-75. Click OK.

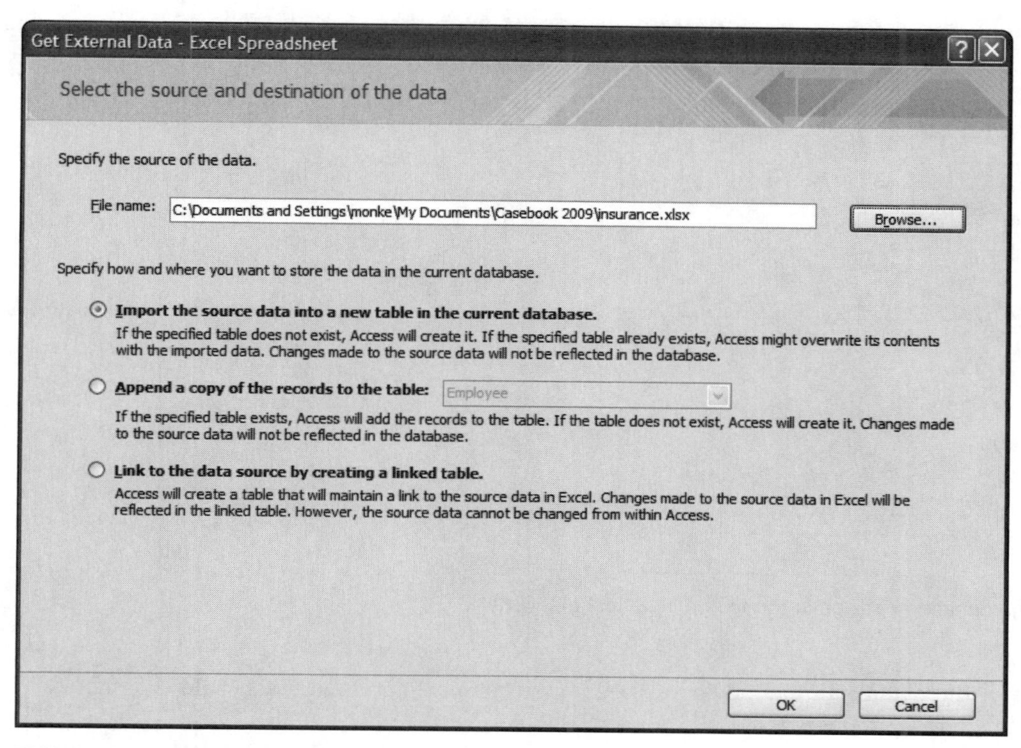

FIGURE B-75 Importing Excel data into a new table

Choose the correct worksheet. Assuming you have just one worksheet in your Excel file, your next screen should look like the one in Figure B-76.

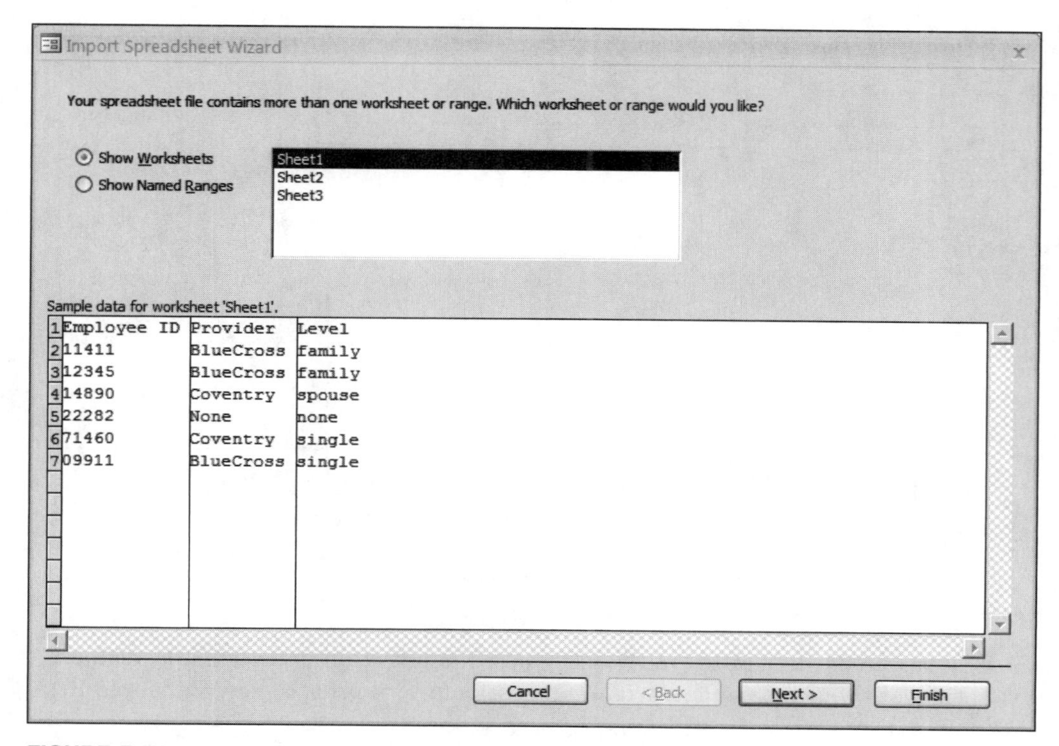

FIGURE B-76 First screen in the Import Spreadsheet Wizard

Choose Next and make sure you select the box that reads First Row Contains Column Headings, as shown in Figure B-77.

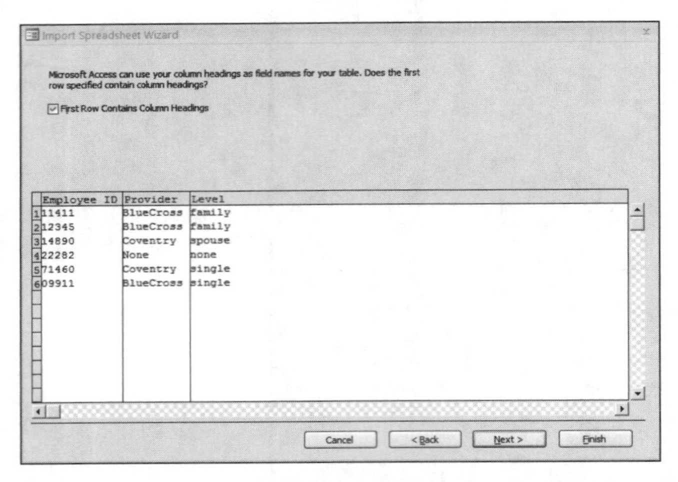

FIGURE B-77 Choosing column headings in the Import Spreadsheet Wizard

Choose Next. Accept the default for each field you are importing on this screen. Each field is assigned a text data type, which is correct for this table. Your screen should look like the one in Figure B-78.

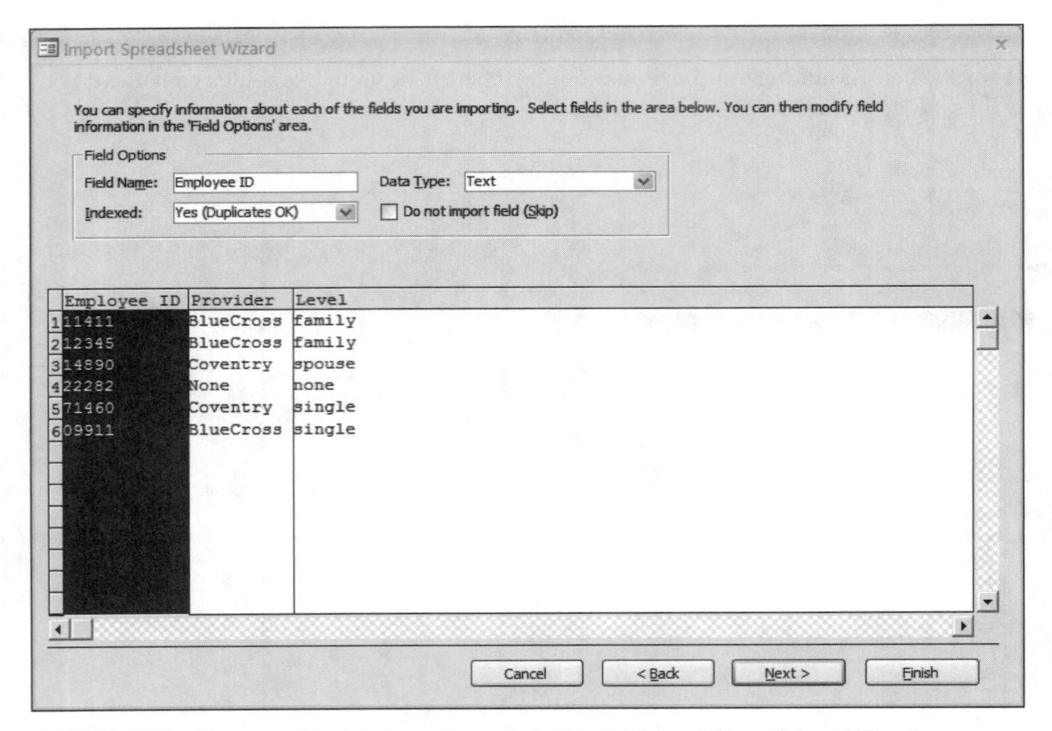

FIGURE B-78 Choosing the data type for each field in the Import Spreadsheet Wizard

Choose Next. In the next screen of the Wizard, you'll be prompted to create an index—that is, define a primary key. Because you will store your data in a new table, choose your own primary key (Employee ID), as shown in Figure B-79.

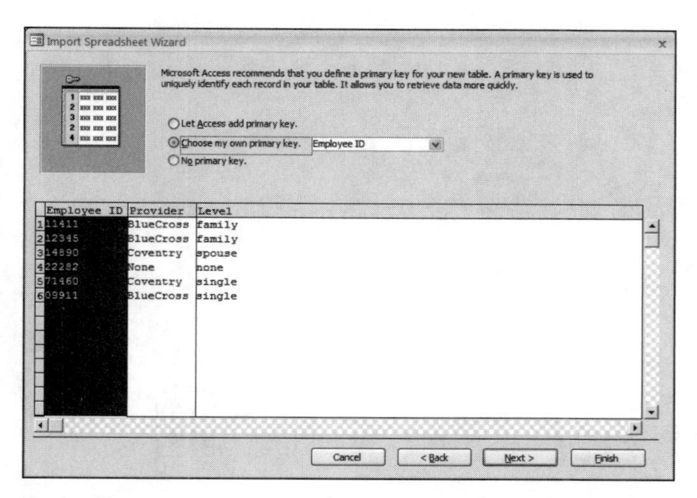

FIGURE B-79 Choosing a primary key field in the Import Spreadsheet Wizard

Continue through the Wizard, giving your table an appropriate name. After importing the table, take a look at it and its design. (Highlight the Table option and use the Design button.) Note the width of each field (very large). Adjust the field properties as needed.

MAKING FORMS

Forms simplify the process of adding new records to a table. Creating forms is easy and can be applied to one or more tables.

When you base a form on one table, you simply select the table and choose the Create tab, Forms Group, Form. The form will then contain all the fields from that table, but only those fields. When data is entered into the form, a complete new record is automatically added to the table. Forms with two tables are discussed next.

Making Forms with Subforms

You also can make a form that contains a subform, which can be useful when you need to create a form that is based on two (or more) tables. Return to the example Employee database to see how forms and subforms would be particularly handy for viewing all the hours that each employee worked each week. Suppose you want to show all of the fields from the EMPLOYEE table; you also want to show the hours each employee worked (in other words, include all fields from the HOURS WORKED table as well).

To create the form and subform, first create a simple one-table form on the EMPLOYEE table. Follow these steps:

1. Select the EMPLOYEE table (click once). In the Create tab, Forms group, choose Form. After the main form is complete, it should resemble the one in Figure B-80.

Employee

Last Name: Brady
First Name: Joseph
Employee ID: 09911
Street Address: 417 Oak Av
City: Elkton
State: MD
Zip: 21921
Date Hired: 9/14/2010
US Citizen: ☑

Record: 1 of 6 No Filter Search

FIGURE B-80 The Employee form

2. To add the subform, choose the HOURS WORKED table from the navigation pane and drag it into the EMPLOYEE form. If you need to stretch the subform to view all fields, click in the subform and drag your cursor across the edge. When a double-tipped arrow appears, hold your mouse button down and drag the edge of the subform to display all fields. Your completed form and subform should resemble that in Figure B-81.

Employee

Employee

Last Name: Brady
First Name: Joseph
Employee ID: 09911
Street Address: 417 Oak Av
City: Elkton
State: MD
Zip: 21921
Date Hired: 9/14/2010
US Citizen: ☑

	Week #	Hours
	1	60
	2	55
*		

Record: 1 of 6 No Filter Search

FIGURE B-81 Form with subform

CREATING A CUSTOM NAVIGATION PANE

If you want someone who knows nothing about Access to run your database, you can create a customized area, known as a navigation pane, to simplify the work. A navigation pane provides a simple, user-friendly interface that has groupings of objects that the user can click to perform certain tasks. For example, you can design a custom navigation pane with two groupings: one for Forms and one for Reports. Your finished product will show the objects within each grouping at the left side of the screen. The user can open each object by clicking it.

To design the custom navigation pane, right-click the top of the navigation pane and choose Navigation Options. The Navigation Options dialog box opens, as shown in Figure B-82.

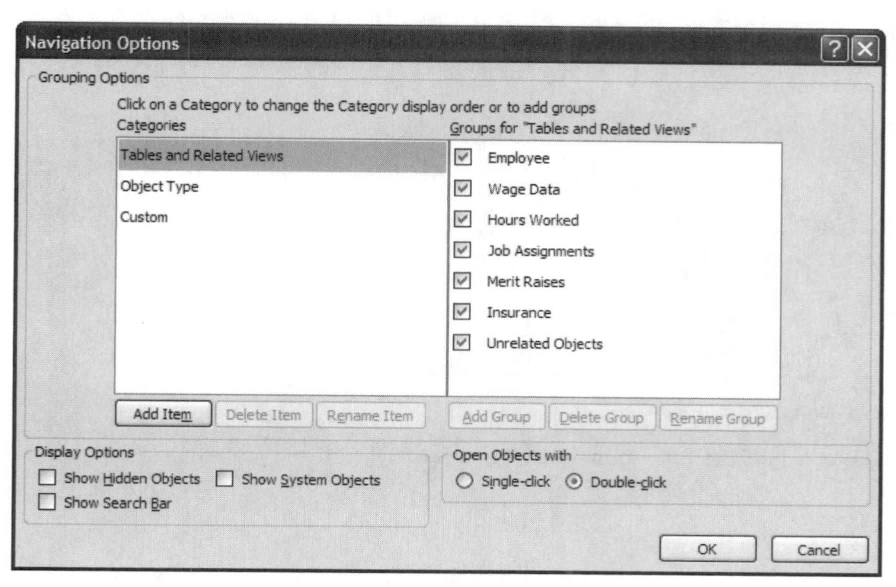

FIGURE B-82 The Navigation Options dialog box

Click the Add Item button, which appears below the Categories column, as shown in Figure B-83.

Navigation Options
Grouping Options
Click on a Category to change the Category display order or to add groups
Categories
Groups for "Easy Navigation"
Tables and Related Views
Object Type
Custom
Easy Navigation
Custom Group 1
Unassigned Objects
Add Item
Delete Item
Rename Item
Add Group
Delete Group
Rename Group
Display Options
Show Hidden Objects
Show System Objects
Show Search Bar
Open Objects with
Single-click
Double-click
OK
Cancel

FIGURE B-83 Add Item option in the Navigation Options dialog box

Change the Custom Group 1 name to Easy Navigation and press Enter.

With Easy Navigation still selected, click the Add Group button, which appears below the right column, Groups for "Easy Navigation." Add a group for Forms and a group for Reports. Leave the Unassigned Objects group as is. Your screen should look like the one shown in Figure B-84.

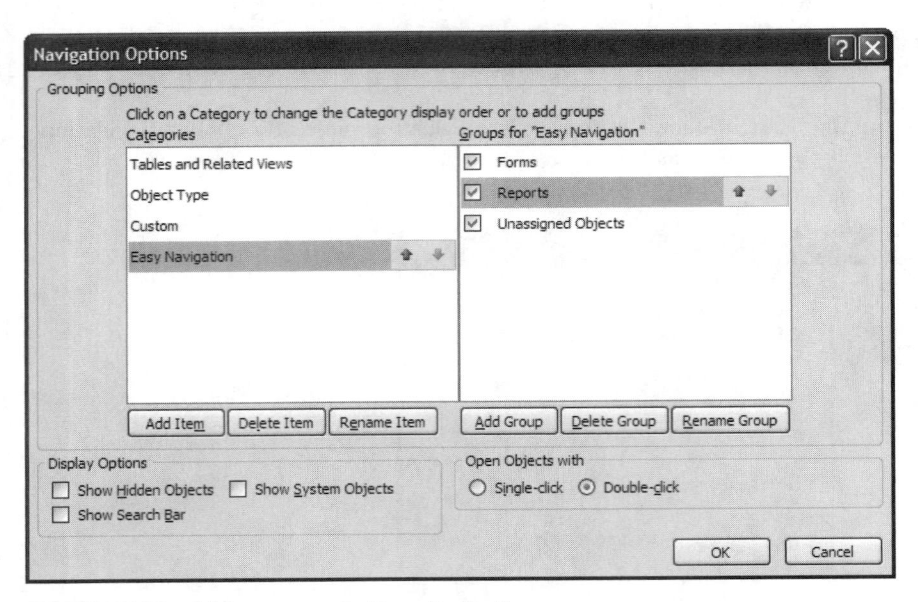

FIGURE B-84 Adding groups to Easy Navigation

Choose OK. Your screen should look like the one shown in Figure B-85.

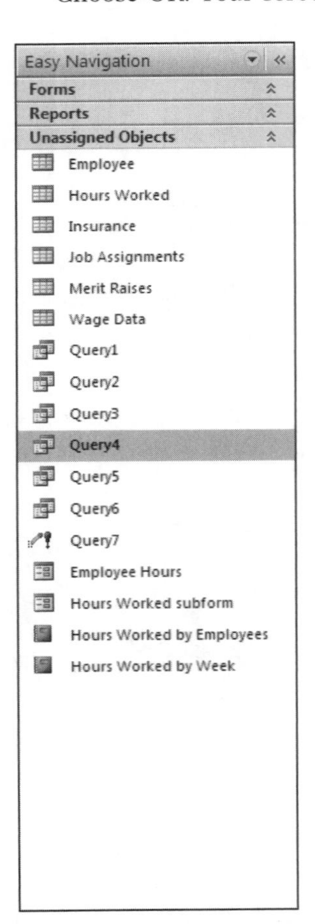

FIGURE B-85 Easy navigation without objects

Now you need to add objects to your new groups. Choose Easy Navigation from the top of the navigation pane.

Drag each form in the Unassigned Objects group to the Forms group and drag each report to the Reports group. When you finish, you can hide the Unassigned Objects group by clicking the double arrow in the Unassigned Objects bar. Your completed Easy navigation pane should look like the one in Figure B-86.

FIGURE B-86 Easy navigation completed

TROUBLESHOOTING COMMON PROBLEMS

Access beginners (and veterans) sometimes create databases that have problems. Some common problems are described below, along with their causes and corrections.

1. *"I saved my database file, but I can't find it on my computer or external secondary storage medium! Where is it?"*

   You saved your file to some fixed disk or some place other than My Documents. Use the Search option of the Windows Start button. Search for all files ending in .accdb (search for *.accdb). If you did save the file, it is on the hard drive (C:\) or on some network drive. (Your site assistant can tell you the drive designators.)

2. *"What is a 'duplicate key field value'? I'm trying to enter records into my Sales table. The first record was for a sale of product X to customer 101, and I was able to enter that one. But when I try to enter a second sale for customer #101, Access tells me I already have a record with that key field value. Am I allowed to enter only one sale per customer?"*

   Your primary key field needs work. You may need a compound primary key—customer number and some other field(s). In this case, customer number, product number, and date of sale might provide a unique combination of values, or you might consider using an invoice number field as a key.

3. *"My query says 'Enter Parameter Value' when I run it. What is that?"*

   This symptom almost always indicates that you have an expression in a Criteria field or calculated field in which *you have misspelled a field name*. Access is very fussy about spelling; for example, it is case-sensitive. Access is also "space-sensitive," meaning that when you insert a space in a field name when defining a table, you must put a space in the field name when you reference it in a query expression. Fix the typo in the query expression.

4. *"I'm getting a fantastic number of rows in my query output—many times more than I need. Most of the rows are duplicates!"*

   This symptom is usually caused by a failure to link all of the tables you brought into the top half of the query generator. The solution is to use the manual click-and-drag method to link the common fields between tables. (Spelling of the field names is irrelevant because the link fields need not be spelled the same.)

5. *"For the most part, my query output is what I expected, but I am getting one or two duplicate rows or not enough rows."*

   You may have linked too many fields between tables. Usually, only a single link is needed between two tables. It's unnecessary to link each common field in all combinations of tables; usually, it's enough to link the primary keys. A layperson's explanation for why overlinking causes problems is that excess linking causes Access to "overthink" the problem and repeat itself in its answer.

   On the other hand, you might be using too many tables in the query design. For example, you brought in a table, linked it on a common field with some other table, but then did not use the table—in other words, you brought down none of its fields and/or you used none of its fields in query expressions. In this case, if you were to get rid of the table, the query would still work. Try doing the following to see whether the few duplicate rows disappear: Click the unneeded table's header in the top of the QBE area and press the Delete key.

6. *"I expected six rows in my query output, but I only got five. What happened to the other one?"*

   Usually, this indicates a data entry error in your tables. When you link the proper tables and fields to make the query, remember that the linking operation joins records from the tables *on common values* (*equal* values in the two tables). For example, if a primary key in one table has the value "123", the primary key or the linking field in the other table should be the same to allow linking. Note that the text string "123" is not the same as the text string "123"—the space in the second string is considered a character too. Access does not see unequal values as an error. Instead, Access moves on to consider the rest of the records in the table for linking. The solution is to look at the values entered into the linked fields in each table and fix any data entry errors.

7. *"I linked fields correctly in a query, but I'm getting the empty set in the output. All I get are the field name headings!"*

   You probably have zero common (equal) values in the linked fields. For example, suppose you are linking on Part Number (which you declared as text). In one field, you have part numbers "001", "002", and "003"; in the other table, you have part numbers "0001", "0002", and "0003". Your tables have no common values, which means no records are selected for output. You'll have to change the values in one of the tables.

8. *"I'm trying to count the number of today's sales orders. A Totals query is called for. Sales are denoted by an invoice number, and I made that a text field in the table design. However, when I ask the Totals query to 'Sum' the number of invoice numbers, Access tells me I cannot add them up! What is the problem?"*

   Text variables are words! You cannot add words, but you can count them. Use the Count Totals operator (not the Sum operator) to count the number of sales, each being denoted by an invoice number.

9. *"I'm doing time arithmetic in a calculated field expression. I subtracted the Time In from the Time Out and got a decimal number! I expected eight hours, and I got the number .33333. Why?"*

   [Time Out] – [Time In] yields the decimal percentage of a 24-hour day. In your case, eight hours is one-third of a day. You must complete the expression by multiplying by 24: ([Time Out] – [Time In]) * 24. Don't forget the parentheses.

10. *"I formatted a calculated field for Currency in the query generator, and the values did show as currency in the query output; however, the report based on the query output does not show the dollar sign in its output. What happened?"*

    Go into the report Design view. A box in one of the panels represents the calculated field's value. Click the box and drag to widen it. That should give Access enough room to show the dollar sign as well as the number in the output.

11. *"I told the Report Wizard to fit all of my output to one page. It does print to one page, but some of the data is missing. What happened?"*

    Access fits all the output on one page by *leaving data out*. If you can stand to see the output on more than one page, deselect the Fit to a Page option in the Wizard. One way to tighten output is to enter Design view and remove space from each box that represents output values and labels. Access usually provides more space than needed.

12. *"I grouped on three fields in the Report Wizard, and the Wizard prints the output in a staircase fashion. I want the grouping fields to be on one line. How can I do that?"*

    Make adjustments in Design view and Layout view. See the Reports section of this tutorial for instructions on making these adjustments.

13. *"When I create an Update query, Access tells me that zero rows are updating or more rows are updating than I want. What is wrong?"*

    If your Update query is not set up correctly (for example, if the tables are not joined properly), Access will try not to update anything or will update all of the records. Check the query, make corrections, and run it again.

14. *"After making a Totals query with a Sum in the Group By row and saving that query, when I go back to it, the Sum field reads 'Expression,' and 'Sum' is entered in the field name box. Is that wrong?"*

    Access sometimes changes that particular statistic when the query is saved. The data remains the same, and you can be assured your query is correct.

15. *"I cannot run my Update query, but I know it is set up correctly. What is wrong?"*

    Check the security content of the database by clicking the Security Content button. You may need to enable certain actions.

CASE 1

PRELIMINARY CASE: SORORITY VOLUNTEER DATABASE

Setting up a Relational Database to Create Tables, Forms, Queries, and Reports

PREVIEW

In this case, you will create a relational database to keep track of a sorority's volunteer work. First, you will create four tables and populate them with data. Next, you will create a form and subform for recording volunteer hours, along with three queries: a Select query, a Totals query, and a query with a calculated field. Finally, you will create a report from a query that displays each volunteer's number of hours.

PREPARATION

- Before attempting this case, you should have some experience using Microsoft Access.
- Complete any part of Access Tutorial B that your instructor assigns, or refer to the tutorial as necessary.

BACKGROUND

The Sigma sorority members at Grand State University in Colorado pride themselves on volunteerism. Over the past 10 years, they have been voted the most philanthropic of all organizations at the school. Every academic year, each woman in the sorority makes a pledge to spend at least 50 hours working in a volunteer activity. The sorority sisters almost see volunteerism as a competition because the sorority member with the most service hours at the end of the year gets to take the sorority's pink convertible car home for the summer!

In the past, the sorority has had some problems recording volunteer hours. For example, some sorority sisters have claimed that not all their hours were written down. Therefore, one sorority member has taken matters into her own hands. After attending an information systems class, she has designed a database to keep track of volunteer hours. You are hired to finish the database project, and you begin the job with the tables already designed for you.

The database contains four tables:

- The MEMBERS table keeps track of member information such as ID, name, e-mail address, and telephone number.
- The CHARITIES table keeps track of all the charities in which the sorority sisters participate. In particular, the table includes charity IDs and names, the contact person at each charity, address, phone number, and e-mail address.
- The EVENTS table records the different charitable events around town. This table includes information about each event, the date, and when the event begins.
- The VOLUNTEER HOURS table records the sorority sisters' hours spent volunteering at specific events.

The sorority members would like to see some requirements for information output in the database beyond simply recording the data. First, they would like to be able to record their volunteer hours. This task can be accomplished with a form and subform.

In addition, the sorority members would like the database to answer some questions. To begin, they would like to see all the events planned for September and October so they can advertise those events as soon as the school year begins.

The sorority likes to see which charities are running the most events. Its women would like a listing of these charities and the number of events announced for the school year so far. Also, the sorority members would like a listing of all charities related to women or girls because some of the members are eager to help that segment of the local population. In addition, the women's club at the university has offered to donate $5 to women's charities for every hour of volunteering. The database will report totals of the sorority sisters' volunteer hours with subtotals of hours spent at particular events. This information should be stored in a nicely formatted report.

ASSIGNMENT 1: CREATING TABLES

Use Microsoft Access to create the tables with the fields shown in Figures 1-1 through 1-4; these tables were discussed in the Background section. Populate the database tables as shown. Add your name to the MEMBERS table, assign yourself a Member ID of 111, and fill in your phone number and e-mail address. Volunteer yourself for at least three events.

Members

Member ID	Member Last Name	Member First Name	Member Email	Member Phone
101	O'Hara	Quinn	ohara@gs.edu	316-559-8451
102	Sanchez	Maria	sanchez@gs.edu	854-954-7895
103	Borgi	Angela	borgi@gs.edu	541-578-6651
104	Amith	Preeti	amith@gs.edu	212-659-8831
105	Jones	Ann	jones@gs.edu	854-773-9654
106	Wanauski	Juliet	wanauski@gs.edu	321-854-9632
107	Watson	Katie	watson@gs.edu	412-965-8745
108	Zeff	Joan	zeff@gs.edu	654-789-8521
109	Bounder	Linda	bounder@gs.edu	212-963-8547
110	Finkel	Marlene	finkel@gs.edu	212-963-8855
*				

FIGURE 1-1 MEMBERS table

Charities

Charity ID	Charity Name	Contact Person	Street Address	City	State	Zip Code	Phone	Charity Email
C-101	Arts Center	Mabel French	101 Oak Avenue	Boulder	CO	80301	484-987-2251	artscenter@comcast.net
C-102	Women's Work	Ann Marie Janusik	59 Maple Lane	Boulder	CO	80329	654-852-7412	ww@gmail.com
C-103	Save the Mountains	Phil Smith	75 Sycamore Street	Boulder	CO	80320	741-854-9654	stm@verizon.net
C-104	Humane Society	George Hangstrom	9132 Pine Way	Boulder	CO	80305	854-996-3651	humane@gmail.com
C-105	Girls' Club	Bobbie Jo Marple	12 Tulip Row	Boulder	CO	80306	714-854-3579	gc@comcast.net
*								

FIGURE 1-2 CHARITIES table

Events

Event ID	Charity ID	Event Name	Date	Start Time
1	C-101	Arts on the Green	10/2/2010	8:00:00 AM
2	C-101	Auction	12/5/2010	8:00:00 PM
3	C-102	Craft Fair	11/26/2010	9:00:00 AM
4	C-103	Trail Cleanup	9/15/2010	8:00:00 AM
5	C-103	Map the Mountains	1/10/2011	10:00:00 AM
6	C-104	Dog Wash	10/15/2010	11:00:00 AM
7	C-105	Pet Photos	12/1/2010	9:00:00 AM
8	C-105	Dog Agility	10/12/2010	12:00:00 PM
9	C-105	Open House	9/1/2010	1:00:00 PM

FIGURE 1-3 EVENTS table

Volunteer Hours

Member ID	Event ID	Number of Hours
101	4	2
101	5	3
101	9	5
102	2	3
102	3	1
102	6	2
103	1	2
104	5	3
104	7	5
104	8	3
105	8	6
106	2	3
106	3	3
106	4	3
107	6	2
108	8	6
108	9	3
109	1	4
109	7	2
109	9	3
110	3	5
110	6	2

FIGURE 1-4 VOLUNTEER HOURS table

ASSIGNMENT 2: CREATING A FORM, QUERIES, AND A REPORT

Assignment 2A: Creating a Form

Create a form so that the sorority can easily record volunteer hours by events. Use the EVENTS table to create the main form. Within this form, create a subform using the VOLUNTEER HOURS table. Save the main form as Events. The subform will be saved within the main form. View one record and, if required by your instructor, print the record. (You should only print one record.) Your output should resemble that in Figure 1-5.

Events

Events

Event ID: 1

Charity ID: C-101

Event Name: Arts on the Green

Date: 10/2/2010

Start Time: 8:00:00 AM

Member ID	Number of Hours
103	2
109	4
*	

Record: 3 of 3 No Filter Search

Record: 1 of 9 No Filter Search

FIGURE 1-5 Events form with subform

Assignment 2B: Creating a Select Query

Create a Select query that lists events in September and October (or any months designated by your instructor). In the output, include columns for Event ID, Event Name, Charity Name, Date, and Start Time. Save the query as September and October Events. Your output should resemble that shown in Figure 1-6.

September and October Events

Event ID	Event Name	Charity Name	Date	Start Time
1	Arts on the Green	Arts Center	10/2/2010	8:00:00 AM
4	Trail Cleanup	Save the Mountains	9/15/2010	8:00:00 AM
6	Dog Wash	Humane Society	10/15/2010	11:00:00 AM
8	Dog Agility	Girls' Club	10/12/2010	12:00:00 PM
9	Open House	Girls' Club	9/1/2010	1:00:00 PM
*				

FIGURE 1-6 September and October Events query

Run the query. Print the results if required.

Assignment 2C: Creating a Totals Query

Create a query that determines the number of events at each charity. Your output should include columns for the Charity Name and the Number of Events. (Note that the Number of Events is a column heading change from the default setting provided by the query generator.) Save your query as Number of Events. Your output should resemble that shown in Figure 1-7. Print the output if desired.

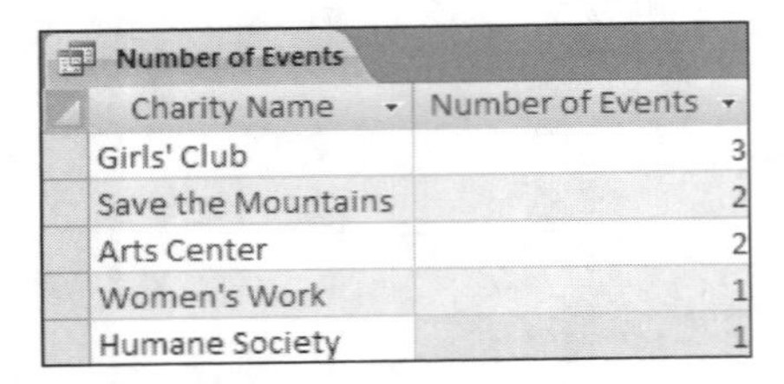

Number of Events

Charity Name	Number of Events
Girls' Club	3
Save the Mountains	2
Arts Center	2
Women's Work	1
Humane Society	1

FIGURE 1-7 Number of Events query

Assignment 2D: Creating a Query with a Calculated Field

Create a query that calculates extra money donated to female charities. Recall that the women's club is willing to donate $5 to women's charities for every hour spent volunteering. The output should include columns for the Charity Name (only for female charities), Contact Person, Phone, Charity Email, and Money Donated, which is a calculated field. (Note that the column heading must be changed from the default setting provided by the query generator.) Save the query as Female Charities. Your output should resemble that shown in Figure 1-8. Print the output if desired.

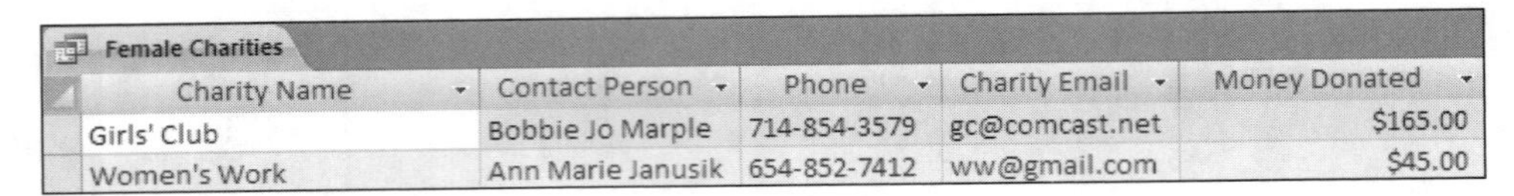

Female Charities

Charity Name	Contact Person	Phone	Charity Email	Money Donated
Girls' Club	Bobbie Jo Marple	714-854-3579	gc@comcast.net	$165.00
Women's Work	Ann Marie Janusik	654-852-7412	ww@gmail.com	$45.00

FIGURE 1-8 Female Charities query

Assignment 2E: Generating a Report

Generate a report based on a query because the data comes from multiple tables. The query should display Member Last Name, Member First Name, Event ID, Event Name, Date, and Number of Hours columns. Based on the query, create a report grouped by Member Last Name. Total the Number of Hours fields to produce a total for each student. Move the Member First Name to the correct level in the report and adjust any output so all fields and data are visible. Save the report as Volunteer Hours. Your output should resemble that in Figure 1-9.

Volunteer Hours

Monday, April 19, 2010
2:34:54 PM

Member Last Name	Member First Name	Event ID	Event Name	Date	Number of Hours
Amith	Preeti				
		5	Map the Mountains	1/10/2011	3
		8	Dog Agility	10/12/2010	3
		7	Pet Photos	12/1/2010	5
					11
Borgi	Angela				
		1	Arts on the Green	10/2/2010	2
					2
Bounder	Linda				
		1	Arts on the Green	10/2/2010	4
		7	Pet Photos	12/1/2010	2
		9	Open House	9/1/2010	3
					9
Finkel	Marlene				
		3	Craft Fair	11/26/2010	5
		6	Dog Wash	10/15/2010	2
					7

FIGURE 1-9 Volunteer Hours report

If you are working with a disk or USB key, make sure you remove it *after* closing the database file.

DELIVERABLES

Assemble the following deliverables for your instructor, either electronically or in printed form:

1. Four tables
2. Form: Events
3. Query 1: September and October Events
4. Query 2: Number of Events
5. Query 3: Female Charities
6. Report: Volunteer Hours
7. Any other required tutorial printouts or electronic media

Staple all pages together. Put your name and class number at the top of each page. If required, make sure that your electronic media are labeled.

CASE **2**

UNIVERSITY SPEEDY DELIVERY

Designing a Relational Database to Create Tables, Forms, Queries, and Reports

PREVIEW

In this case, you will design a relational database for a business that delivers fast food to college students late at night. After your database design is completed and correct, you will create database tables and populate them with data. Then you will produce one form with a subform, four queries, and one report. The queries will answer two questions: What address is associated with a particular order? How many orders are delivered by each delivery person? Another query will list the prices of all items, and the last query will allow the business to increase prices. Your report will display recent orders grouped by customer.

PREPARATION

- Before attempting this case, you should have some experience in database design and in using Microsoft Access.
- Complete any part of Database Design Tutorial A that your instructor assigns.
- Complete any part of Access Tutorial B that your instructor assigns, or refer to the tutorial as necessary.
- Refer to Tutorial F as necessary.

BACKGROUND

The University Speedy Delivery Company, commonly known on campus as USD, delivers good-quality fast food to students around campus late at night. Students can call or text USD to receive items such as pizza, ice cream, and burgers. USD was started by a group of friends who recently graduated. The company is growing so fast that it needs a computerized system to keep track of customer orders.

Here's how the business works: Flyers that list the items available for delivery are distributed to the dorms and apartments around campus. This list is also available on USD's Web site. Before ordering, customers must register on the USD Web site and complete a form that includes their name, local address, cell phone number, e-mail address, and a credit card number. Although USD is strictly a cash business, the credit card is recorded for cases in which customers claim they did not order the food or refuse to pay. Once registered, the student receives a confirmation e-mail and a three-digit customer number. Students can use this customer number to place an order quickly.

Customers can either call in or text their order to USD. When an order comes in, the owner at the office notes the time of day and then takes the order, marking a long preprinted sheet with the items and their prices. The order price is determined by the types of items ordered. The owner is looking forward to the new system, in which he can type the order into a form that will automatically populate the database.

The order is assembled by the owner and delivered by a part-time employee. All delivery people are paid by the hour in cash only. USD expects delivery people to claim this income on their tax returns. Delivery people can also earn hefty tips, especially late at night and in poor weather. Those tips should also be declared as income on tax returns. The rates of pay vary with each delivery person.

After meeting with one of the USD owners, you have several goals for the database once the tables are designed, created, and populated with data. First, you want to produce an easy way for the clerks or owners to input the orders, as described earlier. You propose to develop a prototype system as a form in Microsoft Access that eventually will be migrated to the Web. You've been instructed to create a price list of all the food products for sale. This price list also will be used for advertising purposes.

The owners have complained that some delivery people get orders mixed up. They ask you to develop a method for delivery people to ask the database who ordered a particular item on a particular street. You confidently tell the owners that this method can be accomplished using a query.

The owners would like to track how many orders each delivery person completed in a given time frame. Ideally, they'd like to be able to enter a delivery person's name and then see a number of orders. You suggest a parameter query to accomplish this goal.

The price of food is rising, so the owners want to raise prices by 10% across the board and then have the database raise the prices automatically, rather than having to increase prices manually line by line. You know that an update query will do the job quickly and accurately.

Finally, you need to create a sales report that lists all customers, what they've recently ordered, and the total amount paid. The output of this report will be useful for further marketing the business.

ASSIGNMENT 1: CREATING THE DATABASE DESIGN

In this assignment, you will design your database tables using a word-processing program. Pay close attention to the tables' logic and structure. Do not start developing your Access code in Assignment 2 before getting feedback from your instructor on Assignment 1. Keep in mind that you will need to examine the requirements in Assignment 2 to design your fields and tables properly. It is good programming practice to look at the required outputs before designing your database. When designing the database, observe the following guidelines:

- First, determine the tables you will need by listing the name of each table and the fields it should contain. Avoid data redundancy. Do not create a field if it can be created by a "calculated field" in a query.
- You will need transaction tables. Think about what business events occur with each student's order. Avoid duplicating data.
- Document your tables using the Table feature of your word processor. Your word-processed tables should resemble the format shown in Figure 2-1.
- You must mark the appropriate key field(s) by entering an asterisk (*) next to the field name. Keep in mind that some tables might need a compound primary key to uniquely identify a record within a table.
- Print the database design.

Table Name	
Field Name	Data Type (text, numeric, currency, etc.)
...	...
...	...

FIGURE 2-1 Table design

NOTE

Have your design approved before beginning Assignment 2; otherwise, you may need to redo Assignment 2.

ASSIGNMENT 2: CREATING THE DATABASE, QUERIES, AND REPORT

In this assignment, you will first create database tables in Access and populate them with data. Next, you will create a form with a subform, four queries, and a report.

Assignment 2A: Creating Tables in Access

In this part of the assignment, you will create your tables in Access. Use the following guidelines:

- Enter three types of items—pizza, burgers, and ice cream—and create a realistic price for each type of item.
- Within the three categories, enter three or four flavors or varieties of each item.
- Enter data for at least 10 student customers into the tables. Make up customer IDs, names, addresses, phone numbers, e-mail addresses, and credit card numbers.
- Enter at least two orders per customer.
- Appropriately limit the size of the text fields; for example, a telephone number does not need the default length of 255 characters.
- Print all tables.

Assignment 2B: Creating Forms, Queries, and a Report

You must generate one form, four queries, and one report, as outlined in the Background section of this case.

Form

Create a form and subform based on your ORDERS table and ORDER LINE ITEM table (or whatever you've named these tables). Save the form as ORDERS. Your form should resemble that in Figure 2-2.

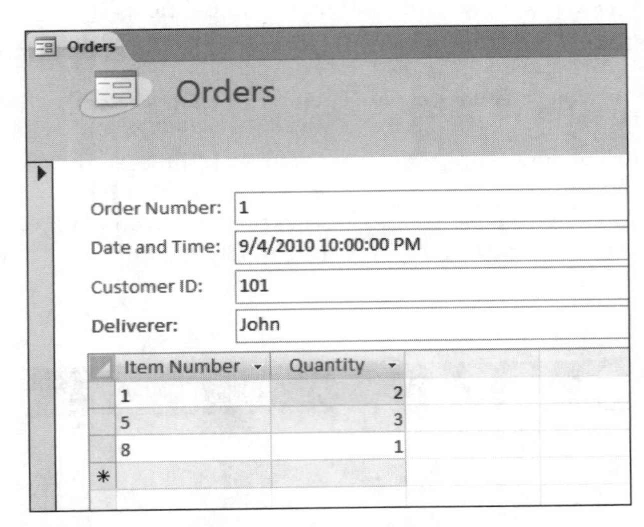

FIGURE 2-2 Orders form and subform

Query 1

Create a query called Price List that lists all the items for sale. The output of the query should include columns for Item Number, Item Type, Flavor, and Price. Your output should resemble that shown in Figure 2-3, although your data will be different.

Price List

Item Number	Item Type	Flavor	Price
7	Burger	Deluxe	$5.00
6	Burger	Cheese	$5.00
5	Burger	Plain	$5.00
11	Ice Cream	Peppermint Stick	$4.50
10	Ice Cream	Vanilla	$4.50
9	Ice Cream	Chocolate	$4.50
8	Ice Cream	Cookie Dough	$4.50
4	Pizza	Veggie	$12.00
3	Pizza	Sausage	$12.00
2	Pizza	Pepperoni	$12.00
1	Pizza	Plain	$12.00

FIGURE 2-3 Price List query

Query 2

Create a query that can report orders for a particular item from a particular street address. For example, the following query is named Ice Cream on Haines. This query would be useful, for example, if a delivery person recalled taking an order for ice cream to be delivered on Haines Street but could not remember the exact address or who ordered the ice cream. Create a query that looks for a delivery of a particular item to a particular street, and that includes columns for Name, Local Address, Cell Phone, Item Type, and Flavor. To find an address on a particular street, consider using a wildcard (*). Your output should look like that in Figure 2-4, although your data will be different.

Ice Cream on Haines

Name	Local Addre:	Cell Phone	Item Type	Flavor
Amit Lun	75 Haines St	212-863-9514	Ice Cream	Cookie Dough

FIGURE 2-4 Ice Cream on Haines query

Query 3

Create a parameter query called Number of Orders Delivered. This query shows an owner how many orders are delivered by each delivery person. The query should prompt the user to enter the name of a delivery person and then calculate the number of orders delivered. Note the column heading change from the default setting provided by the query generator.

Your output should resemble the format shown in Figure 2-5, but the data will be different.

Number of Orders Delivered

Number of Orders Delivered	Deliverer
5	Joe

FIGURE 2-5 Number of Orders Delivered query

Query 4

Create an update query to increase the price of every item by 10%. Run the query to test it. Save the query as Increased Prices.

Report

Create a report named Sales Report. The output should show headings for Name, Local Address, Date and Time, Item Type, and the Total Price, which is a calculated field that sums the total price of all items in the order (price times quantity). First create a query that contains the calculated field and creates input to the report. The report should group data by name and local address. Adjust the output to resemble that shown in Figure 2-6.

Sales Report

Sales Report

Wednesday, April 21, 2010
3:31:17 PM

Name	Local Address	Date and Time	Item Type	Total Price
Amit Lun	75 Haines St			
		9/12/2010 11:15:00 PM	Burger	$10.00
		9/12/2010 11:15:00 PM	Pizza	$12.00
		9/12/2010 11:15:00 PM	Ice Cream	$4.50
				$26.50
Anna Blankenship	515 Cleveland Ave			
		9/4/2010 10:00:00 PM	Pizza	$24.00
		9/4/2010 10:00:00 PM	Burger	$15.00
		9/4/2010 10:00:00 PM	Ice Cream	$4.50
				$43.50
Frank Jordan	45 Orchard Rd			
		9/15/2010 11:00:00 PM	Pizza	$24.00
		9/15/2010 11:00:00 PM	Burger	$10.00
				$34.00

FIGURE 2-6 Sales report

ASSIGNMENT 3: MAKING A PRESENTATION

Create a presentation that explains the database to the owners and delivery staff. Include the design of your database tables and instructions for using the database. Discuss future improvements to the database if the company expands its menu offerings. Your presentation should take fewer than 10 minutes, including a brief question-and-answer period.

DELIVERABLES

Assemble the following deliverables for your instructor, either electronically or in printed form:

1. Word-processed design of tables
2. Tables created in Access
3. Form and subform: Orders
4. Query 1: Price List
5. Query 2: Ice Cream on Haines
6. Query 3: Number of Orders Delivered
7. Query 4: Increased Prices
8. Report: Sales report
9. Presentation materials
10. Any other required tutorial printouts or electronic media

Staple all pages together. Put your name and class number at the top of each page. Make sure that your electronic media are labeled, if required.

CASE 3

THE SUBSTITUTE TEACHER BOOKING DATABASE

Designing a Relational Database to Create Tables, Forms, Queries, and Reports

PREVIEW

In this case, you will design a relational database for a paper-based system that schedules substitute teachers for jobs at schools. After your database design is completed and correct, you will create database tables and populate them with data. Then you will produce one form with a subform, four queries, and one report. The form will allow substitutes to browse the jobs by qualifications. The queries will answer a number of questions: What are the teachers' qualifications? Are any teachers matched with jobs? How many of each type of job is available? How many teachers are qualified for a specified job? Finally, the report will show the duration of each job.

PREPARATION

- Before attempting this case, you should have some experience in database design and in using Microsoft Access.
- Complete any part of Database Design Tutorial A that your instructor assigns.
- Complete any part of Access Tutorial B that your instructor assigns, or refer to the tutorial as necessary.
- Refer to Tutorial F as necessary.

BACKGROUND

Many organizations these days use outsourcing—the subcontracting of business tasks. For example, at some universities, an external company provides food services. The motivation for outsourcing is often not simply to save money, but to eliminate functions that are not central to the organization. For example, a university's core business is education and research, not food service.

In a similar vein, your local school district has decided to outsource its handling of substitute teachers. Coordinating substitute teachers has become so complex that the district doesn't have the clerical staff to handle all the requests. With flu season almost here and with the added threat of swine flu, the district has decided to outsource the business to Substitutes-R-Us, a company that specializes in the scheduling and staffing of substitute teachers.

To create a Web site for substitute teachers to find jobs, Substitutes-R-Us (commonly referred to as SRU) first needs to create a database. SRU interviews the school district clerk who scheduled subs over the last 10 years to find out what information is needed for the database. SRU hires you to attend the interview, listen and gather information, and then design the database.

Joann, the sub clerk, explains what the school district did last year:

Potential subs first registered with the district. They filled out a form that asked for their name, address, phone number, e-mail address, and Social Security number, and they also filled out a qualifications form.

That form basically asked the sub what grades or specialties they were qualified to teach. For example, a sub might be certified to teach K–8th grade, and qualified to teach music. So any given sub might have two or more qualifications we could use. We had one great sub who could teach high school math, special education, and ballroom dancing! But keep in mind that not all teachers are certified. We do hire noncertified teachers, but prefer those who have certification. So it's important that potential subs answer the question about certification on their applications.

We then also compiled a list of schools in the area. That list included the school's name, address, phone number, and principal's name. When a teacher called in sick, the principal would call and ask me to fill the teaching slot with an appropriate substitute. The principal would tell me which grade or special class needed to be taught. For example, a principal from an elementary school would call and ask for a preschool teacher for the next five days. Or, a principal would call and say she needed an art teacher next week because the regular teacher would be out for surgery.

SRU's new system will be based largely on Joann's description of how she scheduled the substitute teachers. SRU will create a Web site where potential substitute teachers can see what jobs are available, and then fill out an online form for the job. Once the job is completed, the information will be sent automatically to the Payroll Department to pay the substitute teacher.

You've been hired as a summer intern because of your experience in database design and your skill with Microsoft Access. Your job is to create a database design for the substitute teacher online booking system and then implement that design.

Specifically, SRU would like you to prepare a prototype form that eventually can be migrated to the Web. The form should allow substitute teachers to look at available jobs by their skill level. SRU would also like you to create a number of queries. For example, it's convenient to have a listing of all the registered subs and their skill level. SRU has also requested a query that matches potential substitute teachers with possible jobs.

Next, SRU would like the database to be able to determine how many jobs are available by skill level. Again, you realize that a query could accomplish this task. SRU's management also would like to be able to enter a qualification level and see how many teachers are qualified at that level. SRU could use this feature for further marketing to schools.

Finally, you need to create a report that lists the duration of each job by school.

ASSIGNMENT 1: CREATING THE DATABASE DESIGN

In this assignment, you will design your database tables using a word-processing program. Pay close attention to the tables' logic and structure. Do not start developing your Access code in Assignment 2 before getting feedback from your instructor on Assignment 1. Keep in mind that you will need to examine the requirements in Assignment 2 to design your fields and tables properly. It is good programming practice to look at the required outputs before designing your database. When designing the database, observe the following guidelines:

- First, determine the tables you will need by listing the name of each table and the fields it should contain. Avoid data redundancy. Do not create a field if it can be created by a "calculated field" in a query.
- You will need a transaction table. Although no money is changing hands in this system, events do occur, and these events are considered transactions. Avoid duplicating data.
- Consider using a logical field when deciding if a teacher is fully qualified.
- Document your tables using the Table feature of your word processor. Your word-processed tables should resemble the format shown in Figure 3-1.
- Mark the appropriate key field(s) by entering an asterisk (*) next to the field name. Keep in mind that some tables might need a compound primary key to uniquely identify a record within a table.
- Print the database design.

Table Name	
Field Name	Data Type (text, numeric, currency, etc.)
...	...
...	...

FIGURE 3-1 Table design

NOTE

Have your design approved before beginning Assignment 2; otherwise, you may need to redo Assignment 2.

ASSIGNMENT 2: CREATING THE DATABASE, QUERIES, AND REPORT

In this assignment, you will first create database tables in Access and populate them with data. Next, you will create a form with a subform, four queries, and a report.

Assignment 2A: Creating Tables in Access

In this part of the assignment, you will create your tables in Access. Use the following guidelines:

- Enter three different schools, each with a different market (for example, high school, elementary school, preschool, etc.)
- Create records for at least 10 substitute teachers. Each should have at least two qualifications.
- Create at least eight jobs with various beginning and ending dates.
- Appropriately limit the size of the text fields; for example, a telephone number does not need the default length of 255 characters.
- Print all tables.

Assignment 2B: Creating Forms, Queries, and a Report

You must generate one form, four queries, and one report, as outlined in the Background section of this case.

Form

Create a form and subform based on your QUALIFICATIONS CODE table and your JOBS table (or whatever you've called these tables). Save the form as Qualifications and Jobs. Your form should resemble that in Figure 3-2.

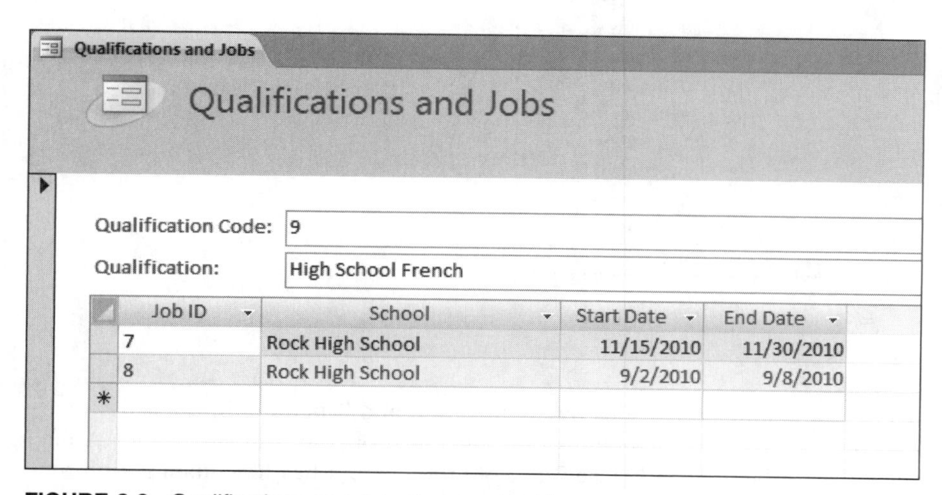

FIGURE 3-2 Qualifications and Jobs form and subform

Query 1

Create a query called Teachers' Qualifications that includes columns for Qualification, Last Name, First Name, Street Address, City, State, Zip, Telephone, and Email address. The query should include all substitute teachers registered at SRU. Your output should resemble that in Figure 3-3, although your data will differ.

Teachers' Qualifications

Qualification	Last Name	First Name	Street Address	City	State	Zip	Telephone	Email address
High School French	Lemole	Sam	4 Fairway Dr	Newark	DE	19711	302-998-7654	sll@comcast.nt
High School Spanish	Lemole	Sam	4 Fairway Dr	Newark	DE	19711	302-998-7654	sll@comcast.nt
K-8	Hewlett	Joanne	57 E. Cherokee St	Wilmington	DE	19803	302-675-2615	jhew@aol.com
9-12	Hewlett	Joanne	57 E. Cherokee St	Wilmington	DE	19803	302-675-2615	jhew@aol.com
Elementary Music	DuBard	Liz	3 Green Valley Way	Landenberg	PA	19350	610-876-3625	Dubard@verizon.ne
Elementary Gym	DuBard	Liz	3 Green Valley Way	Landenberg	PA	19350	610-876-3625	Dubard@verizon.ne
High School Math	Pena	Carlos	30 Peony Ct	Hockessin	DE	19805	302-765-4432	carlos11@zoom.net
High School Spanish	Pena	Carlos	30 Peony Ct	Hockessin	DE	19805	302-765-4432	carlos11@zoom.net
High School French	Wilson	Susan	129 Rockrose Dr	New London	PA	19354	610-873-9287	srw@comcast.net
K-8	Wilson	Susan	129 Rockrose Dr	New London	PA	19354	610-873-9287	srw@comcast.net
9-12	Schoenbeck	Douglas	75 Abbey Lane	Elkton	MD	21921	410-765-3321	schoeny@aol.com
High School Math	Schoenbeck	Douglas	75 Abbey Lane	Elkton	MD	21921	410-765-3321	schoeny@aol.com
Middle School Gym	Seetharam	Ram	115 Marta Dr	Newark	DE	19711	302-775-0011	rams@comcast.net
Middle School Music	Seetharam	Ram	115 Marta Dr	Newark	DE	19711	302-775-0011	rams@comcast.net
Elementary Gym	Klein	Joshua	11 Kress Road	Elkton	MD	21921	410-398-0099	klein@aol.com
Elementary Gym	Jenkins	Amy	95 E Park Pl	Newark	DE	19711	302-737-0098	jenkins12@zoom.ne
Middle School Gym	Dubbs	David	30 Dawes Dr	New London	PA	19354	610-876-4534	dubbs@comcast.ne
High School French	Dubbs	David	30 Dawes Dr	New London	PA	19354	610-876-4534	dubbs@comcast.ne
Middle School Music	Klein	Joshua	11 Kress Road	Elkton	MD	21921	410-398-0099	klein@aol.com
High School French	Jenkins	Amy	95 E Park Pl	Newark	DE	19711	302-737-0098	jenkins12@zoom.ne
*								

FIGURE 3-3 Teachers' Qualifications query

Query 2

Create a query called Job Matching that lists available jobs and the teachers that potentially can fill them, as described in the Background section of this case. The query should include columns for Job ID, School, Start Date, and Last Name of the teacher. Your output will differ, but the layout should resemble that in Figure 3-4.

Job Matching

Job ID	School	Start Date	Last Name
1	Tator Elementary	9/1/2010	Schoenbeck
2	Tator Elementary	9/1/2010	DuBard
1	Tator Elementary	9/1/2010	Pena
2	Tator Elementary	9/1/2010	Jenkins
2	Tator Elementary	9/1/2010	Klein
8	Rock High School	9/2/2010	Lemole
8	Rock High School	9/2/2010	Dubbs
8	Rock High School	9/2/2010	Wilson
8	Rock High School	9/2/2010	Jenkins
5	Rock High School	9/15/2010	Pena
5	Rock High School	9/15/2010	Schoenbeck

FIGURE 3-4 Job Matching query

Query 3

Create a query called Jobs Available that counts all the jobs by qualification and lists them in the order of most available jobs to fewest. The query should include columns for Qualification and Number of Jobs. Your data will differ, but the output should resemble that in Figure 3-5. Note the column heading change from the default setting provided by the query generator.

Jobs Available

Qualification	Number of Jobs
High School Math	2
High School French	2
Middle School Music	1
Middle School Gym	1
High School Spanish	1
Elementary Gym	1

FIGURE 3-5 Jobs Available query

Query 4

Create a query called Number Qualified that prompts for a specified qualification and then lists the number of qualified teachers for that position. The query should include columns for Qualification and Number Qualified data in the output. If you enter High School Math at the prompt, the output will resemble that in Figure 3-6. Note the column heading change from the default setting provided by the query generator.

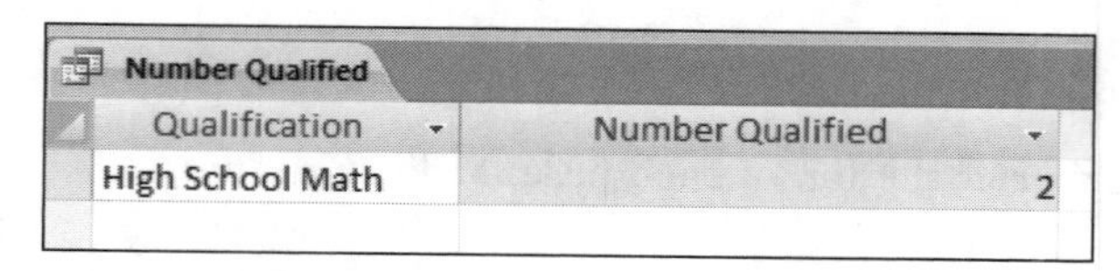
Number Qualified

Qualification	Number Qualified
High School Math	2

FIGURE 3-6 Number Qualified query

Report

Create a report called Length of Jobs by School. To create this report, first create a query that calculates each job's duration (length of time). Include columns for School Name, Qualification, and Duration. Bring the query into the report and group the data by school. Your output should resemble that in Figure 3-7. Make sure that all headings and data are visible.

Length of Jobs by School

School Name	Qualification	Duration
Lakeshore Middle School		
	Middle School Gym	83
	Middle School Music	37
Rock High School		
	High School French	6
	High School French	15
	High School Spanish	7
	High School Math	2
Tator Elementary		
	Elementary Gym	9
	K-8	5

FIGURE 3-7 Length of Jobs by School report

ASSIGNMENT 3: MAKING A PRESENTATION

Create a presentation for the management of SRU and the board of education. Describe your database design and the design of your form, queries, and report. Show samples of the outputs. Your presentation should take fewer than 10 minutes, including a brief question-and-answer period.

DELIVERABLES

Assemble the following deliverables for your instructor, either electronically or in printed form:

1. Word-processed design of tables
2. Tables created in Access
3. Form: Qualifications and Jobs
4. Query 1: Teachers' Qualifications
5. Query 2: Job Matching
6. Query 3: Jobs Available
7. Query 4: Number Qualified
8. Report: Length of Jobs by School
9. Presentation materials
10. Any other required tutorial printouts or electronic media

Staple all pages together. Put your name and class number at the top of each page. Make sure that your electronic media are labeled, if required.

LEASING LUXURY DATABASE

Designing a Relational Database to Create Tables, Forms, Queries, and Reports

PREVIEW

In this case, you will design a relational database for a business that leases designer handbags. After your database design is completed and correct, you will create database tables and populate them with data. Then you will produce one form with a subform, three queries, and two reports. The queries will address the following questions: What bags are available by a specified designer? How many bags from each designer are in inventory? How many days has each customer leased bags? Your reports will display the bags available for rent by designer type and the amount of money each customer has spent leasing different bags.

PREPARATION

- Before attempting this case, you should have some experience in database design and in using Microsoft Access.
- Complete any part of Database Design Tutorial A that your instructor assigns.
- Complete any part of Access Tutorial B that your instructor assigns, or refer to the tutorial as necessary.
- Refer to Tutorial F as necessary.

BACKGROUND

Consumers love luxurious products, but they often cannot afford to buy luxury. Your Aunt Mabel has started a company to address this problem; her new company leases expensive handbags to women. Aunt Mabel owns a large number of designer handbags, and several years ago she began to lend them out to friends for special occasions. After several friends offered to pay her for the "loan" of the handbag, Mabel realized that she could offer a valuable service to women who want an expensive designer handbag for an evening out, a wedding, or other formal event.

Leasing Luxury, or LL as it is commonly known, operates in the Atlanta area. The company model is simple: The daily cost of renting a handbag is set at a certain amount, depending on the designer. For example, a Coach bag rents for $9 per day, shipping included. Customers can keep a bag as long as they like, but they continue to pay the daily fee. Optional insurance costs an extra $1 per day; the insurance covers any damage to the expensive bags.

You have been hired to create a database to keep track of the available handbags, the customers who register for the service, and their bag rentals. The leasing payments are handled manually at this point.

You have several goals for the database. First, you need to keep track of all the handbags that Aunt Mabel owns and that are available for leasing. She categorizes the bags by designer, type of bag, and color. Customer essentials need to be recorded as well. Aunt Mabel conducts the business on paper and via telephone, but she wants to move all work to the computer and Internet as soon as possible. Eventually, she would like customers to book their requests via a Web site.

When customers register for the service, they provide their e-mail address, regular mailing address, and credit card number. (All transactions are conducted with credit cards.) Your database must record this leasing information for the handbags, along with the date rented, date returned, and whether the customer purchases damage insurance.

As customers request handbags, Aunt Mabel and the other workers at LL would like to be able to fill out a form that lists each available bag and includes fields for the customer ID and leasing dates. Mabel would like to be able to enter this information directly into the database. A form with a subform is needed for this purpose. Eventually, this form will be migrated to the Web so that customers can do all their leasing online.

Customers often call and ask what handbags are available by different designers. Mabel would like to be able to enter a designer's name and then see a list of available bags by that designer.

Mabel would like to understand the rental habits of her best customers, who she defines as the customers that keep the bags the longest time. Therefore, she would like to be able to calculate how long customers keep the bags, sorted by the longest duration to the shortest. She would also like to know how many bags she has for each type of designer. This information will be used for marketing purposes.

Finally, you will need to create two reports. The first report lists the bags available by designer and will be used for advertising purposes. The second report will calculate and list the amount of money each customer is spending. The amounts will account for the number of days the bag has been rented and whether the customer purchased the optional insurance.

ASSIGNMENT 1: CREATING THE DATABASE DESIGN

In this assignment, you will design your database tables using a word-processing program. Pay close attention to the tables' logic and structure. Do not start developing your Access code in Assignment 2 before getting feedback from your instructor on Assignment 1. Keep in mind that you will need to examine the requirements in Assignment 2 to design your fields and tables properly. It is good programming practice to look at the required outputs before designing your database. When designing the database, observe the following guidelines:

- First, determine the tables you will need by listing the name of each table and the fields it should contain. Avoid data redundancy. Do not create a field if it can be created by a "calculated field" in a query.
- You will need transaction tables. Think about what business events occur with each customer's actions. Avoid duplicating data.
- Document your tables using the Table feature of your word processor. Your word-processed tables should resemble the format shown in Figure 4-1.
- You must mark the appropriate key field(s) by entering an asterisk (*) next to the field name. Keep in mind that some tables might need a compound primary key to uniquely identify a record within a table.
- Print the database design.

Table Name	
Field Name	Data Type (text, numeric, currency, etc.)
...	...
...	...

FIGURE 4-1 Table design

NOTE

Have your design approved before beginning Assignment 2; otherwise, you may need to redo Assignment 2.

ASSIGNMENT 2: CREATING THE DATABASE, QUERIES, AND REPORTS

In this assignment, you will first create database tables in Access and populate them with data. Next, you will create a form, three queries, and two reports.

Assignment 2A: Creating Tables in Access

In this part of the assignment, you will create your tables in Access. Use the following guidelines:

- Enter at least 10 records for handbags from four different designers.
- Enter records for at least eight customers, including their names, addresses, telephone numbers, e-mail addresses, and fictional credit card numbers. Enter your own name and information as an additional customer.
- Each handbag should be rented at least once. Each customer should rent a handbag twice.
- Appropriately limit the size of the text fields; for example, a telephone number does not need the default length of 255 characters.
- Print all tables if your instructor requires it.

Assignment 2B: Creating Forms, Queries, and Reports

You must generate one form with a subform, three queries, and two reports, as outlined in the Background section of this case.

Form

Create a form and subform based on your HANDBAGS table and RENTALS table (or whatever you've named these tables). Save the form as Handbags. Your form should resemble that in Figure 4-2.

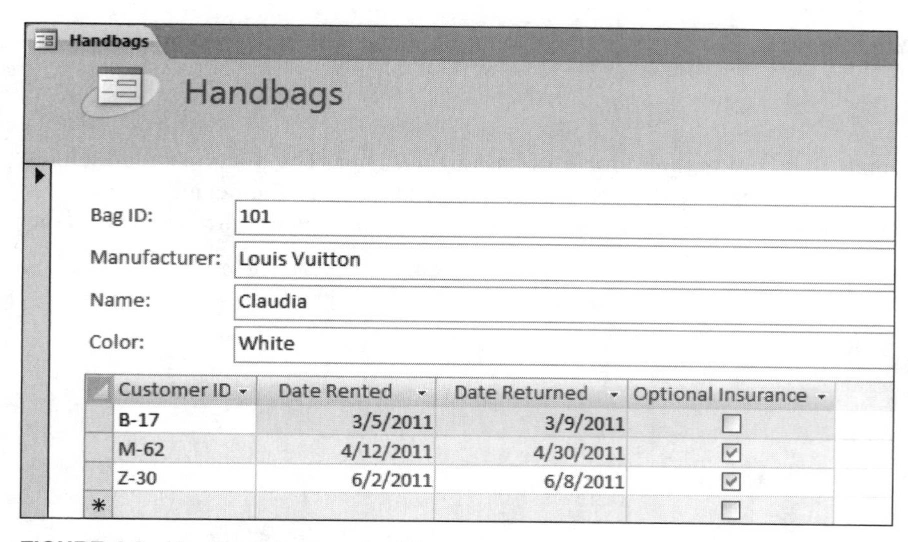

FIGURE 4-2 Handbags form and subform

Query 1

Create a parameter query called Bags by Manufacturer that prompts for the designer name and then lists the name of the bag and its color. In the example shown in Figure 4-3, the manufacturer name is Coach. Your output should resemble that shown in Figure 4-3, although your data will be different.

Bags by Manufacturer

Name	Color	Manufacturer
Satchel	Camel	Coach
Hippie Flap	Green	Coach
Bleecker Bucket	Blue	Coach
*		

FIGURE 4-3 Bags by Manufacturer query

Query 2

Create a query called Best Customers that calculates and adds the rental times of all handbags by each customer. The query should include columns for Last Name, First Name, Address, Telephone, and Total Length of rental. Your output should look like that in Figure 4-4, although your data will be different.

Best Customers

Last Name	First Name	Address	Telephone	Total Length
Zern	Joan	58 W. Central Ave	404-675-0091	40
Murray	Annabelle	59 W. Central Ave	404-998-3928	31
Smith	Patricia	1700 E. Lincoln Ave	404-765-3342	28
Berry	Anna	9 Pleasant Way	404-887-4673	28
Franco	Gina	1012 Peachtree St	404-887-2342	14
Pao	Jill	89 Orchard	404-887-9238	13
Lopato	Maria	5490 West 5th	404-234-8876	7
Quinn	Sally	54 Oak Ave	404-987-3427	1

FIGURE 4-4 Best Customers query

Query 3

Create a query called Inventory Totals that counts the number of bags by designer. The query should include columns for Manufacturer and Number of Bags. Note the column heading change from the default setting provided by the query generator. Your output should resemble the format shown in Figure 4-5, but your data will be different.

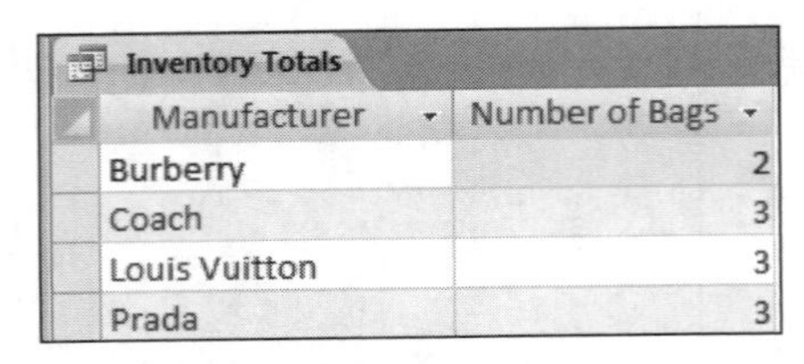

Inventory Totals

Manufacturer	Number of Bags
Burberry	2
Coach	3
Louis Vuitton	3
Prada	3

FIGURE 4-5 Inventory Totals query

Report 1

Create a report named Handbags. The report's output should show headings for Manufacturer, Bag ID, Name, and Color. Group the report by Manufacturer. Add an appropriate piece of clip art to the upper-left corner of the report. (Use the program's Help feature for assistance.) Depending on your data, your output should resemble that shown in Figure 4-6.

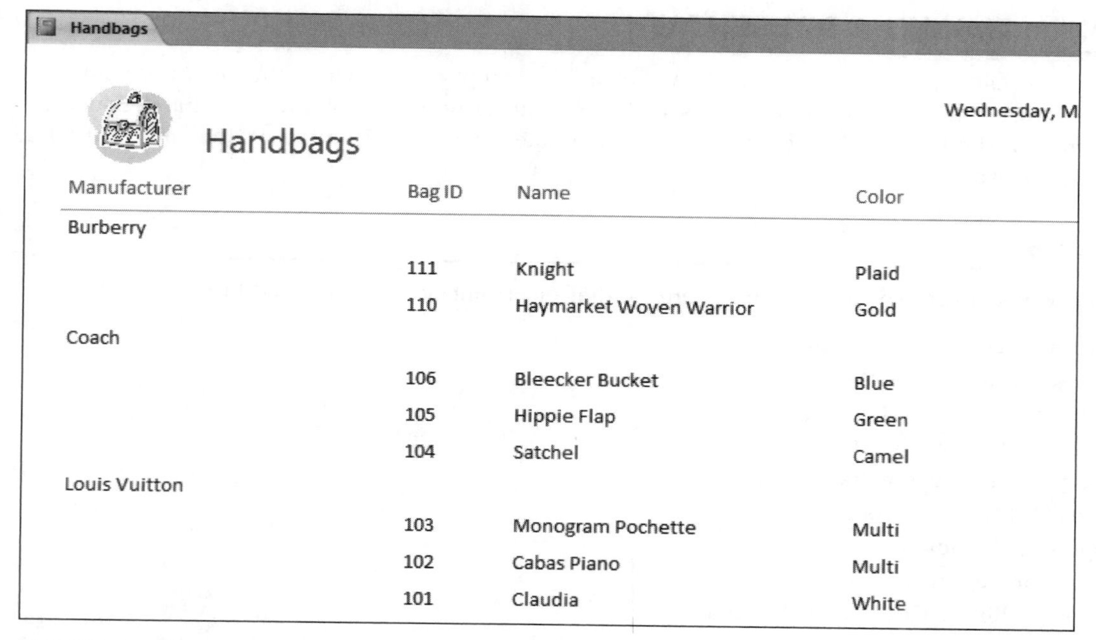

Handbags

Handbags

Wednesday, M

Manufacturer	Bag ID	Name	Color
Burberry			
	111	Knight	Plaid
	110	Haymarket Woven Warrior	Gold
Coach			
	106	Bleecker Bucket	Blue
	105	Hippie Flap	Green
	104	Satchel	Camel
Louis Vuitton			
	103	Monogram Pochette	Multi
	102	Cabas Piano	Multi
	101	Claudia	White

FIGURE 4-6 Handbags report

Report 2

Create a report called Rental Costs per Customer that includes columns for Last Name, First Name, Manufacturer, Name, and Cost of each rental. You will first need to create a query for this report. The query will have a calculated field to determine the cost of each rental. Use an If statement in the calculated field to add any insurance costs that the customer might incur. Group the report by the customer's name, and subtotal each group's cost of rental. Depending on your data, your report should resemble that in Figure 4-7.

Rental Costs per Customer

Wednesday, May 05, 2010
3:01:51 PM

Last Name	First Name	Manufacturer	Name	Cost
Berry	Anna			
		Louis Vuitton	Monogram Pochette	$175.00
		Louis Vuitton	Claudia	$35.00
		Coach	Bleecker Bucket	$36.00
				$246.00
Franco	Gina			
		Louis Vuitton	Cabas Piano	$78.00
		Coach	Hippie Flap	$40.00
		Coach	Satchel	$20.00
				$138.00

FIGURE 4-7 Rental Costs per Customer report

ASSIGNMENT 3: MAKING A PRESENTATION

Create a presentation that explains the database to Mabel and her employees. Include the design of your database tables and instructions for using the database. Discuss future improvements to the database such as tracking customers' favorite rentals. Your presentation should take fewer than 10 minutes, including a brief question-and-answer period.

DELIVERABLES

Assemble the following deliverables for your instructor, either electronically or in printed form:

1. Word-processed design of tables
2. Tables created in Access
3. Form and subform: Handbags
4. Query 1: Bags by Manufacturer
5. Query 2: Best Customers
6. Query 3: Inventory Totals
7. Report 1: Handbags
8. Query for Report 2
9. Report 2: Rental Costs per Customer
10. Presentation materials
11. Any other required tutorial printouts or electronic media

Staple all pages together. Put your name and class number at the top of each page. Make sure that your electronic media are labeled, if required.

CASE **5**

THE CONFERENCE CENTER RESERVATION DATABASE

Designing a Relational Database to Create Tables, Forms, Queries, Reports, and Navigation Panes

PREVIEW

In this case, you will design a reservation database for a conference center. After your database design is completed and correct, you will create database tables and populate them with data. Then you will produce a form with a subform, six queries, a report, and a navigation pane. The form and subform will record all the bookings for each conference room. The queries will display customers in the banking industry, the equipment packages ordered with each conference room rental, the prices for food orders, the length of each reservation, the cost of any equipment rental, and the cost of any food required. The report will summarize the total cost to each customer. The navigation pane will allow access to the form and subform, queries, and report.

PREPARATION

- Before attempting this case, you should have some experience in database design and in using Microsoft Access.
- Complete any part of Database Design Tutorial A that your instructor assigns.
- Complete any part of Access Tutorial B that your instructor assigns, or refer to the tutorial as necessary.
- Refer to Tutorial F as necessary.

BACKGROUND

A conference center in the rolling hills of Pennsylvania has been bought by a new owner who is attempting to revive the business. The conference rooms have been renovated and are now open for booking. The new owner, Bud Michels, is a friend of your uncle. At a recent golf outing, Bud mentioned that he needed help in setting up a computerized reservation system for the conference center. Your uncle told him that you were working on database design and implementation in your university studies. Bud decided to hire you as a summer intern to help him create a system for reserving conference rooms.

The first step is to design the database; you and Bud need to discuss the business so you can understand how the database tables should be set up. The conference center is a three-story building on the site of a botanical garden. The center has seven rooms available for booking; the smallest room holds 25 people and the largest holds 800. Each room is named after a flower commonly found in the gardens. Clients pay a per-person fee depending on how many participants attend a conference session. Bud thinks this is a smart pricing strategy. He tells you that each participant needs a chair and table, as well as water, a candy dish, and paper and pen. His costs rise as the number of session participants increases. Charges are also levied for rental of audiovisual equipment and microphones, if requested.

When a client calls to reserve a conference room, a clerk collects information such as company name, contact person, telephone number, and e-mail address. Each client is assigned a unique ID number, so there is no confusion among clients with similar names. When clients book conference rooms, they are required to specify the meeting dates and times and how many people will attend. Clients are offered a number of options for their meetings. The "AV package" includes an overhead projector with a hookup to a computer, DVD player, or video player. The "microphone package" activates a wireless sound system that works with a microphone.

The conference rooms come with five meal packages. The basic package includes no meals but offers water and candy dishes. Three other packages offer continual coffee and tea, breakfast, or lunch. The comprehensive package offers a combination of continual coffee and tea, breakfast, and lunch.

After the database is designed and the tables are created and populated, Bud will need information from the database. First, he would like to be able to enter a room number and see all reservations for that conference room. You decide that the best way to handle this request is to create a form and subform.

Bud would like the database to answer a number of questions. You explain that queries within the database are the perfect solution. First, he'd like to know which clients are in the banking industry so he can market to them. A second query would list all ordered equipment packages by date and floor of the conference center; this query would be useful for the workers who set up the rooms.

Another required query must calculate the cost of food packages based on the number of participants in a conference. An additional query needs to calculate the duration of each conference at the center. Two other queries need to calculate the cost of renting equipment and the cost of food for each client.

Bud would like you to create a report that facilitates further business analysis. The report should display the total cost of the conference for each client.

Finally, you offer to create a navigation pane to simplify access to the form and subform, queries, and report.

ASSIGNMENT 1: CREATING THE DATABASE DESIGN

In this assignment, you will design your database tables using a word-processing program. Pay close attention to the tables' logic and structure. Do not start developing your Access code in Assignment 2 before getting feedback from your instructor on Assignment 1. Keep in mind that you will need to examine the requirements in Assignment 2 to design your fields and tables properly. It's good programming practice to look at the required outputs before designing your database. When designing the database, observe the following guidelines:

- First, determine the tables you will need by listing the name of each table and the fields it should contain. Avoid data redundancy. Do not create a field if it can be created by a "calculated field" in a query.
- You will need a number of transaction tables. Avoid duplicating data.
- Document your tables using the Table feature of your word processor. Your word-processed tables should resemble the format shown in Figure 5-1.
- You must mark the appropriate key field(s) by entering an asterisk (*) next to the field name. Keep in mind that some tables might need a compound primary key to uniquely identify a record within a table.
- Print the database design.

Table Name	
Field Name	Data Type (text, numeric, currency, etc.)
...	...
...	...

FIGURE 5-1 Table design

NOTE

Have your design approved before beginning Assignment 2; otherwise, you may need to redo Assignment 2.

ASSIGNMENT 2: CREATING THE DATABASE, QUERIES, REPORT, AND NAVIGATION PANE

In this assignment, you will first create database tables in Access and populate them with data. Next, you will create a form with a subform, six queries, a report, and a navigation pane.

Assignment 2A: Creating Tables in Access

In this part of the assignment, you will create your tables in Access. Use the following guidelines:

- Enter records for at least five clients in the tables. Make up fictional company names and other details.
- Assume that two types of equipment packages are available for rental: the AV package, which costs $150 per day; and the microphone package, which costs $50 per day.
- Clients can choose from five types of food packages: the basic package, at $10 per person; continual coffee and tea for $15 per person; lunch for $20 per person; breakfast and lunch for $30 per person; or the comprehensive package (a combination of continual coffee and tea, breakfast, and lunch) for $40 per person.
- Create a booking for each client. Make some bookings longer than one day; have these clients order equipment and food.
- Name each room after a flower or plant; create data for at least five rooms.
- Appropriately limit the size of the text fields; for example, a telephone number does not need the default length of 255 characters.
- Print all tables if your instructor requires it.

Assignment 2B: Creating Forms, Queries, Report, and Navigation Pane

You will create one form with a subform, six queries, one report, and one navigation pane, as outlined in the Background section of this case.

Form

Create a reservation form that includes all the information about the conference rooms. Include the details of each room reservation as the subform. Your data will vary but the output should resemble that in Figure 5-2.

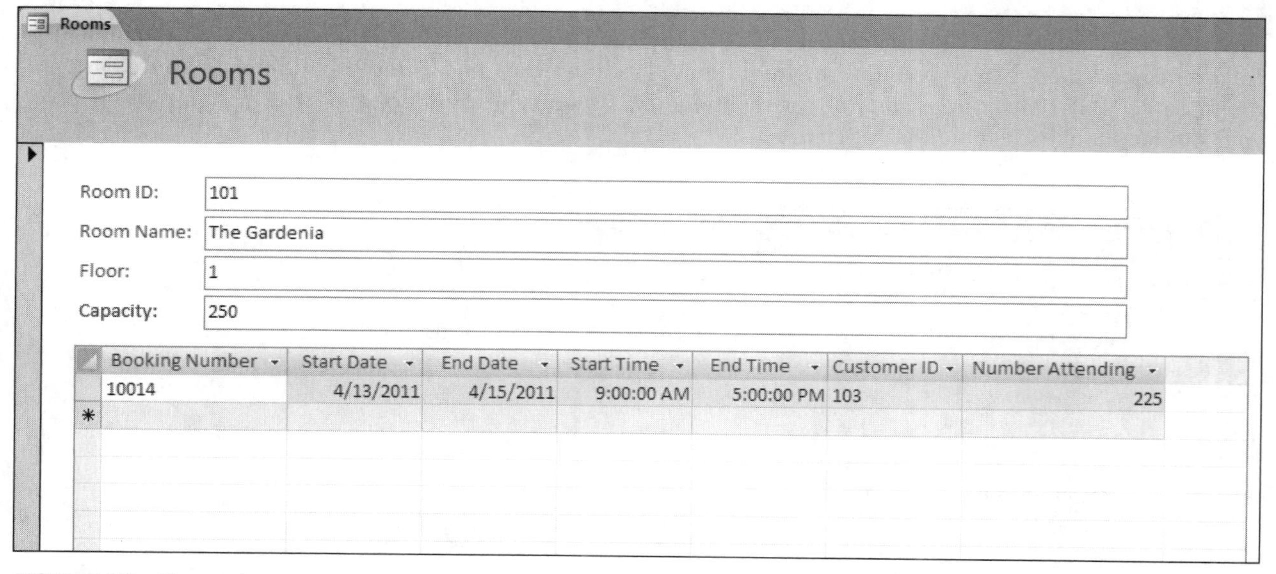

FIGURE 5-2 Rooms form

Query 1

Create a query called Banking Customers that only lists clients who are involved in the banking industry. Display headings for Customer Name, Contact Person, Customer Phone, and Customer Email. To filter client data appropriately, consider using a wildcard to display any customers with the word "bank" in their records. Your data will differ, but your output should resemble that in Figure 5-3.

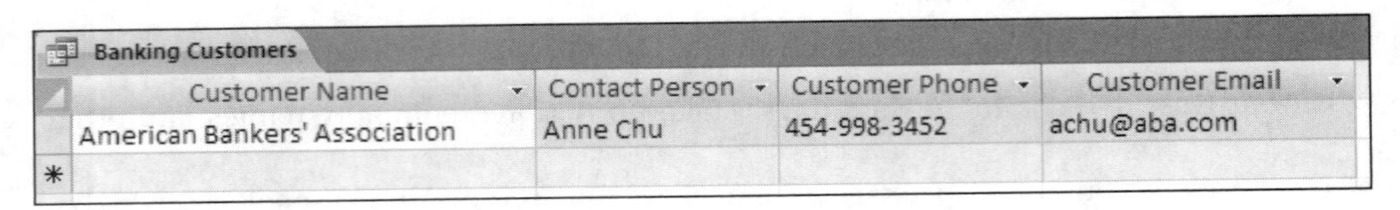

Banking Customers

Customer Name	Contact Person	Customer Phone	Customer Email
American Bankers' Association	Anne Chu	454-998-3452	achu@aba.com
*			

FIGURE 5-3 Banking Customers query

Query 2

Create a query called Equipment Packages Required. Display headings for Room Name, Start Date, Floor, Item Name, and Number Attending. To better assist workers in setting up the rooms, sort the output by date and then by floor. Your data will differ, but the output should resemble that in Figure 5-4. Note the column heading change from the default setting provided by the query generator.

Equipment Packages Required

Room Name	Start Date	Floor	Item Name	Number Attending
The Tulip	3/2/2011	1	AV Equipment	50
The Orchid	3/30/2011	3	AV Equipment	500
The Gardenia	4/13/2011	1	AV Equipment	225
The Rose	4/15/2011	1	Microphone	115
The Rose	4/15/2011	1	AV Equipment	115
The Tulip	5/1/2011	1	Microphone	20
The Tulip	5/1/2011	1	AV Equipment	20
*				

FIGURE 5-4 Equipment Packages Required query

Query 3

Create a query called Food Price Packages. This query should prompt for the number of people attending the conference and then display columns for Item Name, Price Per Day, and Total Price, which is a calculation of the Price Per Day times the number of people attending. If you run this query and enter data for 100 people, your output should resemble that in Figure 5-5.

Food Price Packages

Item Name	Price Per Day	Total Price
Basic Per Person	$10.00	$1,000.00
Lunch Per Person	$20.00	$2,000.00
Breakfast/lunch Per Person	$30.00	$3,000.00
Coffee/Tea Per Person	$15.00	$1,500.00
Full Package Per Person	$40.00	$4,000.00
*		

FIGURE 5-5 Food Price Packages query

Query 4

Create a query called Booking Duration that lists the Room ID, Room Name, and Number of Days the room is reserved. Display the rooms in order of most days booked to least. If you run this query, your output should resemble that in Figure 5-6.

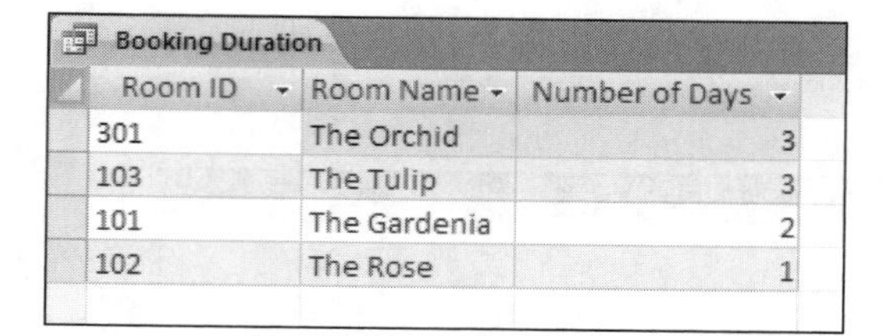

Booking Duration

Room ID	Room Name	Number of Days
301	The Orchid	3
103	The Tulip	3
101	The Gardenia	2
102	The Rose	1

FIGURE 5-6 Booking Duration query

Query 5

Create a query called Equipment Costs. Display headings for Customer Name, Equipment Costs (calculated field), Item Name, and Start Date. Your data will differ, but the output should resemble that in Figure 5-7.

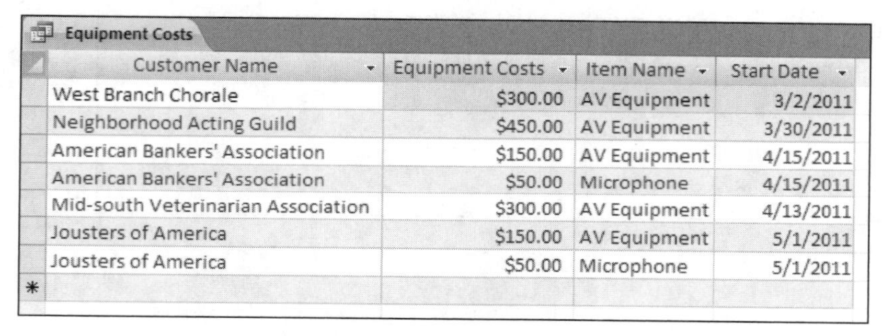

Equipment Costs

Customer Name	Equipment Costs	Item Name	Start Date
West Branch Chorale	$300.00	AV Equipment	3/2/2011
Neighborhood Acting Guild	$450.00	AV Equipment	3/30/2011
American Bankers' Association	$150.00	AV Equipment	4/15/2011
American Bankers' Association	$50.00	Microphone	4/15/2011
Mid-south Veterinarian Association	$300.00	AV Equipment	4/13/2011
Jousters of America	$150.00	AV Equipment	5/1/2011
Jousters of America	$50.00	Microphone	5/1/2011

FIGURE 5-7 Equipment Costs query

Query 6

Create a query called Cost of Food. Display headings for Customer Name, Start Date, Cost of Food (calculated field), and Item Name. Your data will differ, but the output should resemble that in Figure 5-8.

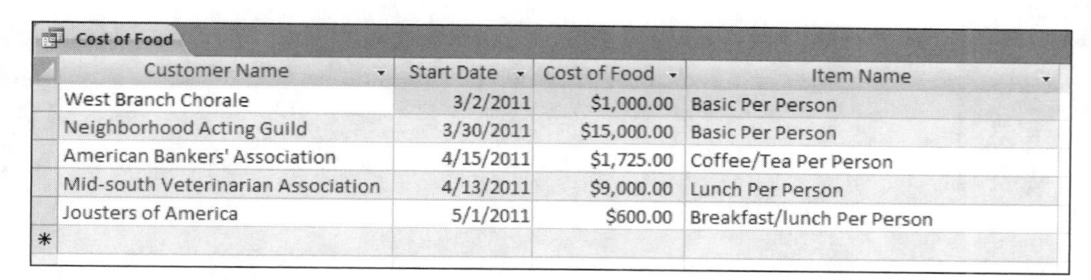

Cost of Food

Customer Name	Start Date	Cost of Food	Item Name
West Branch Chorale	3/2/2011	$1,000.00	Basic Per Person
Neighborhood Acting Guild	3/30/2011	$15,000.00	Basic Per Person
American Bankers' Association	4/15/2011	$1,725.00	Coffee/Tea Per Person
Mid-south Veterinarian Association	4/13/2011	$9,000.00	Lunch Per Person
Jousters of America	5/1/2011	$600.00	Breakfast/lunch Per Person

FIGURE 5-8 Cost of Food query

Report

Create a report called Final Bill. Your report's output should include headings for Customer Name, Contact Person, Customer Phone, Start Date, and Total Bill. You must first create a query as a basis for this report. Use Queries 5 and 6 (described previously) as input to the query for this report. Your data will differ, but your output should resemble that in Figure 5-9.

Final Bill

Final Bill

Friday, April 30, 2010
3:43:10 PM

Customer Name	Contact Person	Customer Phone	Start Date	Total Bill
American Bankers' Association	Anne Chu	454-998-3452	4/15/2011	$1,775.00
American Bankers' Association	Anne Chu	454-998-3452	4/15/2011	$1,875.00
West Branch Chorale	Daniel Sanchez	324-983-2876	3/2/2011	$1,300.00
Mid-south Veterinarian Association	Susan Morehead	765-879-0011	4/13/2011	$9,300.00
Jousters of America	Philip Samuel	321-980-7765	5/1/2011	$650.00
Jousters of America	Philip Samuel	321-980-7765	5/1/2011	$750.00
Neighborhood Acting Guild	John Smith	512-651-2233	3/30/2011	$15,450.00

Page 1 of 1

FIGURE 5-9 Final Bill report

Navigation Pane

Create a navigation pane called Easy Navigation to access the form, queries, and reports. Your navigation pane should look like that in Figure 5-10.

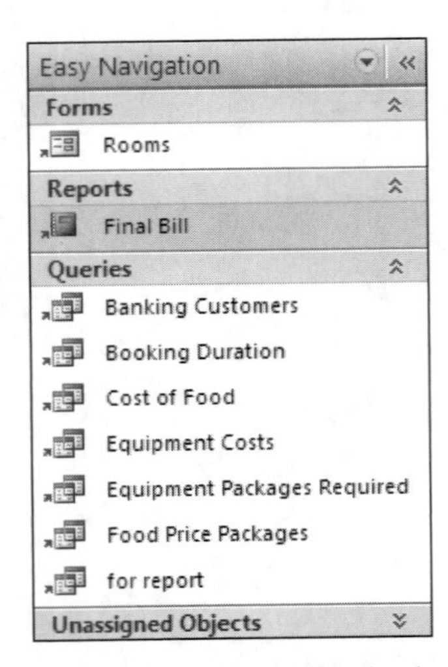

FIGURE 5-10 Easy Navigation pane

ASSIGNMENT 3: MAKING A PRESENTATION

Create a presentation for Bud and his workers. Pay particular attention to potential database users who are not familiar with Microsoft Access. Your presentation should take fewer than 15 minutes, including a brief question-and-answer period.

DELIVERABLES

Assemble the following deliverables for your instructor, either electronically or in printed form:

1. Word-processed design of tables
2. Tables created in Access
3. Form: Rooms
4. Query 1: Banking Customers
5. Query 2: Equipment Packages Required
6. Query 3: Food Price Packages
7. Query 4: Booking Duration
8. Query 5: Equipment Costs
9. Query 6: Cost of Food
10. Report: Final Bill
11. Navigation Pane: Easy Navigation (not printed)
12. Any other required tutorial printouts or electronic media

Staple all pages together. Put your name and class number at the top of each page. Make sure that your electronic media are labeled, if required.

PART 2

DECISION SUPPORT CASES USING EXCEL SCENARIO MANAGER

BUILDING A DECISION SUPPORT SYSTEM IN EXCEL

A **decision support system (DSS)** is a computer program that can represent, either mathematically or symbolically, a problem that a user needs to solve. Such a representation is, in effect, a model of a problem.

Here's how a DSS program works: The DSS program accepts input from the user or looks at data in files on a disk. Then the DSS program runs the input and any other necessary data through the model. The program's output is the information the user needs to solve a problem. Some DSS programs recommend a solution to a problem.

A DSS can be written in any programming language that lets a programmer represent a problem. For example, a DSS can be built in a third-generation language such as Visual Basic or in a database package such as Access. A DSS also can be written in a spreadsheet package such as Excel.

The Excel spreadsheet package has standard built-in arithmetic functions as well as many statistical and financial functions. Thus, many kinds of problems—such as those in accounting, operations, and finance—can be modeled in Excel.

This tutorial has the following four sections:

1. **Spreadsheet and DSS Basics**—In this section, you'll learn how to create a DSS program in Excel. Your program will be in the form of a cash flow model. This section will give you practice in spreadsheet design and in building a DSS program.
2. **Scenario Manager**—In this section, you'll learn how to use an Excel tool called the Scenario Manager. With any DSS package, one problem with playing "what if" is: Where do you physically record the results from running each set of data? Typically, a user writes the inputs and related results on a sheet of paper. Then—ridiculously enough—the user might have to input the data *back* into a spreadsheet for further analysis. The Scenario Manager solves that problem. It can be set up to capture inputs and results as "scenarios," which are then summarized on a separate sheet in the Excel workbook.
3. **Practice Using Scenario Manager**—You will work on a new problem, a case using the Scenario Manager.
4. **Review of Excel Basics**—This brief section reviews the information you'll need to do the spreadsheet cases that follow this tutorial.

SPREADSHEET AND DSS BASICS

Assume it is late in Year 1 of a three-year period, and you are trying to build a model of a company's net income (profit) and cash flow for the next two years (Year 2 and Year 3). The problem is trying to forecast net income and cash flow in those years. The company is likely to use the forecasts to answer a business question or make a strategic decision, so the estimates should be as accurate as you can make them. After researching the problem, you decide that the estimates should be based on three factors: (1) Year 1 results, (2) estimates of the underlying economy, and (3) the cost of products the company sells.

Your model will use an income statement and cash flow framework. The user can input values for two possible states of the economy in Year 2–Year 3: an O for an optimistic outlook or a P for a pessimistic outlook. The state of the economy is expected to affect the number of units the company can sell as well as each unit's selling price. In a good, or O, economy, more units can be sold at a higher price. The user also can input values into your model for two possible price trends in the cost of goods sold: a U for up or a D for down. A U means that the cost of an item sold will be higher than it was in Year 1; a D means that the cost will be less.

Presumably, the company will do better in a good economy and with lower input costs—but how much better? Such relationships are too complex for most people to assess in their head, but a software model can assess them easily. Thus, the user can play "what if" with the input variables and note the effect on net income and year-end cash levels. For example, a user can ask these questions: What if the economy is good and costs go up? What will net income and cash flow be in that case? What will happen if the economy is down and costs go down? What will be the company's net income and cash flow in that case? With an Excel software model available, the answers to those questions are easily quantified.

Organization of the DSS Model

Your spreadsheets should have the following sections:

- Constants
- Inputs
- Summary of Key Results
- Calculations (of values that will be used in the Income and Cash Flow Statements)
- Income and Cash Flow Statements

Here, as an extended illustration, a DSS model is built for the forecasting problem described. Next, you'll look at each spreadsheet section. Figure C-1 and Figure C-2 show how to set up the spreadsheet.

	A	B	C	D
1	**Tutorial Exercise**			
2				
3	**Constants**	**Year 1**	**Year 2**	**Year 3**
4	Tax Rate	NA	33%	35%
5	Number of Business Days	NA	300	300
6				
7	**Inputs**	**Year 1**	**Year 2**	**Year 3**
8	Economic Outlook (O=Optimistic, P=Pessimistic)	NA		NA
9	Purchase Price Outlook (U=Up, D=Down)	NA		NA
10				
11	**Summary of Key Results**	**Year 1**	**Year 2**	**Year 3**
12	Net Income after Taxes	NA		
13	End-of-year cash on hand	NA		
14				
15	**Calculations**	**Year 1**	**Year 2**	**Year 3**
16	Number of units sold in a day	1000		
17	Selling Price per unit	$7.00		
18	Cost of goods sold per unit	$3.00		
19	Number of units sold in a year	NA		

FIGURE C-1 Tutorial skeleton 1

	A	B	C	D
21	**Income and Cash Flow Statements**	**Year 1**	**Year 2**	**Year 3**
22	Beginning-of-year cash on hand	NA		
23	Sales (Revenue)	NA		
24	Cost of goods sold	NA		
25	Income before taxes	NA		
26	Income tax expense	NA		
27	Net income after taxes	NA		
28	End-of-year cash on hand	$10,000		

FIGURE C-2 Tutorial skeleton 2

Each spreadsheet section is discussed next.

NOTE

Some of the numbers in the spreadsheet shown here have no decimals, whereas some have two decimals. Decimal precision can be controlled using the decimal icons in the Home tab's Number group. That formatting procedure is discussed more fully in the "Review of Excel Basics" section of this tutorial.

The Constants Section

This section of Figure C-1 records the values used in spreadsheet calculations. In a sense, the constants are inputs, except that they do not change. In this tutorial, constants are Tax Rate and the Number of Business Days.

The Inputs Section

The inputs shown in Figure C-1 are for the Economic Outlook and Purchase Price Outlook (manufacturing input costs). Inputs could conceivably be entered for *each year* the model is covered (here, Year 2 and Year 3), which would let you enter an O for Year 2's economy in one cell and a P for Year 3's economy in another cell. Alternatively, one input for the two-year period could be entered in one cell. For simplicity, this tutorial uses the latter approach.

The Summary of Key Results Section

This section of the spreadsheet captures Year 2 and Year 3 Net Income after Taxes (profit) and End-of-year cash on hand, which you should assume are the two relevant outputs of this model. The summary merely repeats results in one easy-to-see place; otherwise, these results would appear in widely spaced places in the spreadsheet. (It also makes for easier charting later.)

The Calculations Section

This area is used to compute the following data:

- The Number of units sold in a day, which is a function of the Year 1 value and the economic outlook input
- The Selling Price per unit, which is similarly derived
- The Cost of goods sold per unit, which is a function of the Year 1 value and of purchase-price outlook
- The Number of units sold in a year, which equals the number of units sold in a day times the number of business days in a year

Those formulas could be embedded in the Income and Cash Flow Statements section of the spreadsheet, which will be described shortly. Doing that, however, would result in expressions that are complex and difficult to understand. Putting the intermediate calculations into a separate Calculations section breaks up the work into modules. This is good form because it simplifies your programming.

The Income and Cash Flow Statements Section

This section is the "body" of the spreadsheet. It shows the following:

- Beginning-of-year cash on hand, which equals cash at the end of the *prior* year.
- Sales (Revenue), which equals the units sold in the year times the unit selling price.
- Cost of goods sold, which is units sold in the year times the price paid to acquire or make the unit sold.
- Income before taxes, which equals sales, less total costs.
- Income tax expense, which is zero when there are losses; otherwise, it is the income before taxes times the tax rate. (Income tax expense is sometimes called Income taxes.)
- Net income after taxes, which equals income before taxes less income tax expense.
- End-of-year cash on hand, which is beginning-of-year cash on hand plus net income. (In the real world, cash flow estimates must account for changes in receivables and payables. In this case, assume that sales are collected immediately—that is, there are no receivables or bad debts. Also assume that suppliers are paid immediately—that is, there are no payables.)

Construction of the Spreadsheet Model

Next, you will work through the following three steps to build your spreadsheet model:

1. Make a "skeleton" of the spreadsheet and call it **TUTC.xlsx**.
2. Fill in the "easy" cell formulas.
3. Enter the "hard" spreadsheet formulas.

Make a Skeleton

Your first step is to set up a skeleton worksheet. The worksheet should have headings, text string labels, and constants—but no formulas.

To set up the skeleton, you must first grasp the problem conceptually. The best way to do that is to work backward from what the "body" of the spreadsheet will look like. Here, the body is the Income and Cash Flow Statements section. Set up the body in your mind or on paper, and then do the following:

- Decide what amounts should be in the Calculations section. In the income statement of this tutorial's model, Sales (revenue) will be Number of units sold in a day times Selling Price per unit. You will calculate the intermediate amounts (Number of units sold in a year and Selling Price per unit) in the Calculations section.
- Set up the Summary of Key Results section by deciding what *outputs* are needed to solve the problem. The Inputs section should be reserved for amounts that can change: the controlling variables, which are the Economic Outlook and the Purchase Price Outlook.
- Use the Constants section for values you will need to use but that are not in doubt; that is, you will not have to input them or calculate them. Here, the Tax Rate is a good example of such a value.

AT THE KEYBOARD

Type in the Excel skeleton shown in Figure C-1 and Figure C-2.

NOTE

A designation of NA means that a cell will not be used in any formula in the worksheet. The Year 1 values are needed only for certain calculations; so for the most part, the Year 1 column's cells show NA. (Recall that the forecast is for Year 2 and Year 3.) Also be aware that you can "break" a text string in a cell by pressing the Alt and Enter keys at the same time at the break point, which makes the cell "taller." To show centered data and create borders in cells, see "Formatting Cells" later in this tutorial.

Fill in the "Easy" Formulas

The next step in building a spreadsheet model is to fill in the "easy" formulas. To prepare, you should format the cells in the Summary of Key Results section so they contain no decimals. (For details, see "Formatting Cells" later in this tutorial.) As previously mentioned, the Summary of Key Results section (see Figure C-3) simply echoes results shown in other places. Consider Figures C-1 and C-2, and note that C27 in Figure C-2 will hold the Net Income after Taxes. You need to echo that amount in C12 of Figure C-1, so the formula in C12 is =C27. The simple translation is: "Copy what is in C27 into C12."

NOTE

With the insertion point in C12, the cell's contents—in this case, the formula =C27—appear in the editing window above the lettered column indicators, as shown in Figure C-3.

C12	=C27		
	B	C	D
11 Summary of Key Results	Year 1	Year 2	Year 3
12 Net Income after Taxes	NA	$0	
13 End-of-year cash on hand	NA		

FIGURE C-3 Echo Year 2 Net Income after Taxes

At this point, C27 is empty (and thus has a zero value), but that does not prevent you from copying. Copy cell C12's formula to the right, to cell D12. Copying puts =D27 into D12, which is what you want. (Year 3's Net Income after Taxes is in D27.)

To perform the Copy operation, use the following steps:

1. Click in the cell or range of cells that you want to copy.
2. Hold down the Control key and press C (Ctrl+C).
3. Select the destination cell. (If a range of cells is the destination, select the upper-left cell of the destination range.)
4. Hold down the Control key and press V (Ctrl+V).
5. Press the Escape key to deactivate the copied cell (or range).

You can also take the following steps to copy cell contents:

1. Select the Home tab.
2. Click in the cell or range of cells that you want to copy.
3. In the Clipboard group, select Copy.
4. Select the destination cell. (If a range of cells is the destination, select the upper-left cell of the destination range.)
5. In the Clipboard group, select Paste.
6. Press the Escape key to deactivate the copied cell (or range).

As you can see in Figure C-4, End-of-year cash on hand for Year 2 cash is in cell C13. Echo the cash results in cell C28 to cell C13. (Put =C28 in cell C13, as shown in Figure C-4.) Copy the formula from C13 to D13.

C13	=C28		
	B	C	D
11 Summary of Key Results	Year 1	Year 2	Year 3
12 Net Income after Taxes	NA	$0	$0
13 End-of-year cash on hand	NA	$0	

FIGURE C-4 Echo Year 2 End-of-year cash on hand

At this point, the Calculations section formulas will not be entered because they are not all "easy" formulas. Move on to the easier formulas in the Income and Cash Flow Statements section, as if the calculations were already done. Again, the fact that the Calculations section cells are empty does not stop you from entering formulas in this section. You should format the cells in the Income and Cash Flow Statements section for zero decimals.

As you can see in Figure C-5, Beginning-of-year cash on hand is the cash on hand at the end of the *prior* year. In C22 for Year 2, type =B28. The "skeleton" you just entered is shown in Figure C-5. Cell B28 has the End-of-year cash on hand for Year 1.

C22	=B28			
	A	B	C	D
21	**Income and Cash Flow Statements**	**Year 1**	**Year 2**	**Year 3**
22	Beginning-of-year cash on hand	NA	$10,000	
23	Sales (Revenue)	NA		
24	Cost of goods sold	NA		
25	Income before taxes	NA		
26	Income tax expense	NA		
27	Net income after taxes	NA		
28	End-of-year cash on hand	$10,000		

FIGURE C-5 Echo of End-of-year cash on hand for Year 1 to Beginning-of-year cash on hand for Year 2

Figure C-6 shows the next step, which is to copy the formula in cell C22 to the right. Sales (Revenue) is the Number of units sold in a year times Selling Price per unit. In cell C23, enter =C17*C19, as shown in Figure C-6.

C23	=C17*C19			
	A	B	C	D
15	**Calculations**	**Year 1**	**Year 2**	**Year 3**
16	Number of units sold in a day	1000		
17	Selling Price per unit	$7.00		
18	Cost of goods sold per unit	$3.00		
19	Number of units sold in a year	NA		
20				
21	**Income and Cash Flow Statements**	**Year 1**	**Year 2**	**Year 3**
22	Beginning-of-year cash on hand	NA	$10,000	$0
23	Sales (Revenue)	NA	$0	
24	Cost of goods sold	NA		
25	Income before taxes	NA		
26	Income tax expense	NA		
27	Net income after taxes	NA		
28	End-of-year cash on hand	$10,000		

FIGURE C-6 Enter the formula to compute Year 2 sales

The formula C17*C19 multiplies the unit selling price by the units sold for the year. (Cells C17 and C19 are empty now, which is why Sales shows a zero after the formula is entered.) Copy the formula to the right, to D23.

The Cost of goods sold is handled similarly. In C24, type =C18*C19, which equals Cost of goods sold per unit times Number of units sold in a year. Copy the formula to the right.

In cell C25, the formula for Income before taxes is =C23–C24. Enter the formula and copy it to the right.

In the United States, income taxes are paid only on positive income before taxes. In cell C26, the Income tax expense is zero when the Income before taxes is zero or less; otherwise, Income tax expense equals the Tax Rate times the income before taxes. The Tax Rate is a constant (in C4). An IF statement is needed to express this logic:

IF(Income before taxes is <= 0,
  then put zero tax in C26,
  else, in C26, put a number equal to multiplying the Tax Rate by the Income before taxes)

C25 stands for Income before taxes, and C4 stands for Tax Rate. In Excel, substitute those cell addresses:

=IF(C25 <= 0, 0, C4 * C25)

Copy the income tax expense formula to the right.

In cell C27, Net income after taxes is Income before taxes less Income tax expense: =C25–C26. Enter and copy the formula to the right.

The End-of-year cash on hand is the Beginning-of-year cash on hand plus Net income after taxes. In cell C28, enter =C22+C27. The Income and Cash Flow Statements section at that point is shown in Figure C-7. Copy the formula to the right.

C28 | =C22+C27

	A	B	C	D
21	**Income and Cash Flow Statements**	Year 1	Year 2	Year 3
22	Beginning-of-year cash on hand	NA	$10,000	$10,000
23	Sales (Revenue)	NA	$0	$0
24	Cost of goods sold	NA	$0	$0
25	Income before taxes	NA	$0	$0
26	Income tax expense	NA	$0	$0
27	Net income after taxes	NA	$0	$0
28	End-of-year cash on hand	$10,000	$10,000	

FIGURE C-7 Status of Income and Cash Flow Statements

Put in the "Hard" Formulas

The next step is to finish the spreadsheet by filling in the "hard" formulas.

AT THE KEYBOARD

In C8, enter an O for Optimistic, and in C9, enter U for Up. There is nothing magical about these values—they just give the worksheet formulas some input to process. Recall that the inputs will cover both Year 2 and Year 3. Enter NA in D8 and D9 to remind yourself that those cells will not be used for input or by other worksheet formulas. Your Inputs section should look like the one shown in Figure C-8.

	A	B	C	D
7	**Inputs**	Year 1	Year 2	Year 3
8	Economic Outlook (O=Optimistic, P=Pessimistic)	NA	O	NA
9	Purchase Price Outlook (U=Up, D=Down)	NA	U	NA

FIGURE C-8 Entering two input values

Recall that cell addresses in the Calculations section are already referred to in formulas in the Income and Cash Flow Statements section. The next step is to enter formulas for those calculations. Before doing that, format the Number of units sold in a day and Number of units sold in a year for zero decimals, and format the Selling Price per unit and Cost of goods sold per unit for two decimals.

The easiest formula in the Calculations section is the Number of units sold in a year, which is the Number of Business Days (in C5) times the Number of units sold in a day (in C16). In C19, enter =C5*C16, as shown in Figure C-9.

C19 | =C5*C16

	A	B	C	D
1	**Tutorial Exercise**			
2				
3	**Constants**	Year 1	Year 2	Year 3
4	Tax Rate	NA	33%	35%
5	Number of Business Days	NA	300	300
6				
7	**Inputs**	Year 1	Year 2	Year 3
8	Economic Outlook (O=Optimistic, P=Pessimistic)	NA	O	NA
9	Purchase Price Outlook (U=Up, D=Down)	NA	U	NA
10				
11	**Summary of Key Results**	Year 1	Year 2	Year 3
12	Net Income after Taxes	NA	$0	$0
13	End-of-year cash on hand	NA	$10,000	
14				
15	**Calculations**	Year 1	Year 2	Year 3
16	Number of units sold in a day	1000		
17	Selling Price per unit	$7.00		
18	Cost of goods sold per unit	$3.00		
19	Number of units sold in a year	NA	0	

FIGURE C-9 Entering the formula to compute Year 2 Number of units sold in a year

Copy the formula to cell D19 for Year 3.

Assume that if the Economic Outlook is Optimistic, the Year 2 Number of units sold in a day will be 6% more than in Year 1; in Year 3, they will be 6% more than in Year 2. Also assume that if the Economic Outlook is Pessimistic, the Number of units sold in a day in Year 2 will be 1 percent less than those sold in Year 1; in Year 3, they will be 1 percent less than those sold in Year 2. An IF statement is needed in C16 to express this idea:

IF(economy variable = Optimistic,
then Number of units sold in a day will go UP 6%,
else Number of units sold in a day will go DOWN 1%)

Substituting cell addresses:

=IF($C$8 = "O", B16 * 106%, B16 * 99%)

NOTE

In Excel, quotation marks denote text. The input is one letter of text, so the quotation marks around the O are needed. Also note that multiplying by 106% results in a 6% increase, whereas multiplying by 99% results in a 1% decrease.

Enter the entire IF formula into cell C16, as shown in Figure C-10. Absolute addressing is needed ($C$8) because the address is in a formula that gets copied *and* you do not want the cell reference to change (to D8, which has the value NA) when you copy the formula to the right. Absolute addressing maintains the $C$8 reference when the formula is copied. Copy the formula in C16 to D16 for Year 3.

C16 =IF($C$8="O",B16*106%,B16*99%)

	A	B	C	D
15	**Calculations**	**Year 1**	**Year 2**	**Year 3**
16	Number of units sold in a day	1000	1060	
17	Selling Price per unit	$7.00		
18	Cost of goods sold per unit	$3.00		

FIGURE C-10 Entering the formula to compute Year 2 Number of units sold in a day

The Selling Price per unit is also a function of the Economic Outlook. Assume that the two-part rule is as follows:

- If the Economic Outlook is Optimistic, the Selling Price per unit in Year 2 will be 1.07 times that of Year 1; in Year 3, it will be 1.07 times that of Year 2.
- On the other hand, if the Economic Outlook is Pessimistic, the Selling Price per unit in Year 2 and Year 3 will equal the per-unit price in Year 1; that is, the price will not change.

Test your understanding of the selling price calculation by figuring out the formula for cell C17. Enter the formula and copy it to the right. You will need to use absolute addressing. (Can you see why?)

The Cost of goods sold per unit is a function of the Purchase Price Outlook. Assume that the two-part rule is as follows:

- If the Purchase Price Outlook is Up (U), Cost of goods sold per unit in Year 2 will be 1.25 times that of Year 1; in Year 3, it will be 1.25 times that of Year 2.
- On the other hand, if the Purchase Price Outlook is Down (D), the multiplier each year will be 1.01.

Again, to test your understanding, figure out the formula for cell C18. Enter and copy the formula to the right. You will need to use absolute addressing.

Your formulas for selling price and cost of goods sold, given Optimistic and Up input values, should yield the calculated values shown in Figure C-11.

	A	B	C	D
15	**Calculations**	**Year 1**	**Year 2**	**Year 3**
16	Number of units sold in a day	1000	1060	1124
17	Selling Price per unit	$7.00	$7.49	$8.01
18	Cost of goods sold per unit	$3.00	$3.75	$4.69
19	Number of units sold in a year	NA	318,000	337,080

FIGURE C-11 Calculated values given Optimistic and Up input values

Assume that you change the input values to Pessimistic and Down. Your formulas should yield the calculated values shown in Figure C-12.

	A	B	C	D
15	**Calculations**	**Year 1**	**Year 2**	**Year 3**
16	Number of units sold in a day	1000	990	980
17	Selling Price per unit	$7.00	$7.00	$7.00
18	Cost of goods sold per unit	$3.00	$3.03	$3.06
19	Number of units sold in a year	NA	297,000	294,030

FIGURE C-12 Calculated values given Pessimistic and Down input values

That completes the body of your spreadsheet. The values in the Calculations section ripple through the Income and Cash Flow Statements section because the income statement formulas reference the calculations. Assuming inputs of Optimistic and Up, the income and cash flow numbers should look like those in Figure C-13.

	A	B	C	D
21	**Income and Cash Flow Statements**	**Year 1**	**Year 2**	**Year 3**
22	Beginning-of-year cash on hand	NA	$10,000	$806,844
23	Sales (Revenue)	NA	$2,381,820	$2,701,460
24	Cost of goods sold	NA	$1,192,500	$1,580,063
25	Income before taxes	NA	$1,189,320	$1,121,398
26	Income tax expense	NA	$392,476	$392,489
27	Net income after taxes	NA	$796,844	$728,909
28	End-of-year cash on hand	$10,000	$806,844	$1,535,753

FIGURE C-13 Completed Income and Cash Flow Statements section

SCENARIO MANAGER

You are now ready to use the Scenario Manager to capture inputs and results as you play "what if" with the spreadsheet.

There are four possible combinations of input values: O-U (Optimistic-Up), O-D (Optimistic-Down), P-U (Pessimistic-Up), and P-D (Pessimistic-Down). Financial results for each combination will be different. Each combination of input values can be referred to as a "scenario." Excel's Scenario Manager records the results of each combination of input values as a separate scenario and then shows a summary of all scenarios in a separate worksheet. Those summary worksheet values can be used as a raw table of numbers and then printed or copied into a Microsoft Word document. The table of data can then be the basis for an Excel chart, which also can be printed or inserted into a memo.

You have four possible scenarios for the economy and the purchase price of goods sold: Optimistic-Up, Optimistic-Down, Pessimistic-Up, and Pessimistic-Down. The four input-value sets produce different financial results. When you use the Scenario Manager, define the four scenarios; then have Excel (1) sequentially run the input values "behind the scenes" and (2) put the results for each input scenario in a summary sheet.

When you define a scenario, you give it a name and identify the input cells and input values. Then you identify the output cells so Excel can capture the outputs in a summary sheet.

AT THE KEYBOARD

To start, select the Data tab. The Data Tools group has a What-If Analysis icon. Select the drop-down arrow and click the Scenario Manager menu choice. Initially, no scenarios are defined, as you can see in Figure C-14.

FIGURE C-14 Initial Scenario Manager window

You can use this window to add, delete, or edit scenarios. Toward the end of the process, you also can create the summary sheet.

NOTE

When working with this window and its successors, do not press the Enter key to navigate. Use mouse clicks to move from one step to the next.

To continue defining a scenario, click the Add button. In the resulting Add Scenario window, name the first scenario Opt-Up. Then type the input cells in the Changing cells field—here, they are C8:C9. (Note that C8 and C9 are contiguous input cells. Noncontiguous input cell ranges are separated by a comma.) Excel may add dollar signs to the cell address, but do not be concerned about that. The window should look like the one shown in Figure C-15.

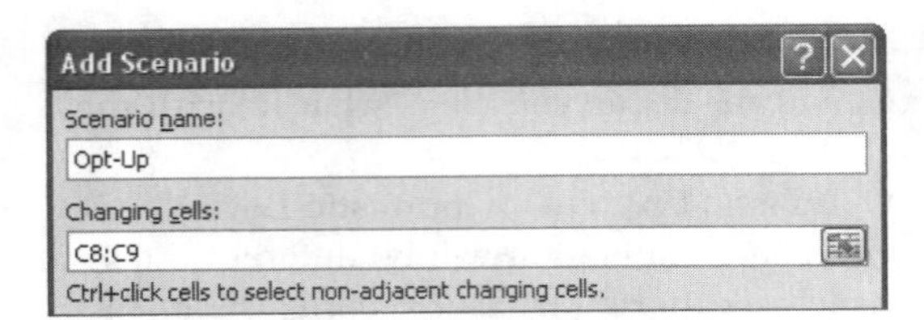

FIGURE C-15 Entering Opt-Up as a scenario

Now click OK, which moves you to the Scenario Values window. Here you indicate what the input *values* will be for the scenario. The values in the *current* spreadsheet cells will be displayed. They might or might not be correct for your scenario. For the Opt-Up scenario, you need to enter an O and a U, if not the current values. Enter those values if needed, as shown in Figure C-16.

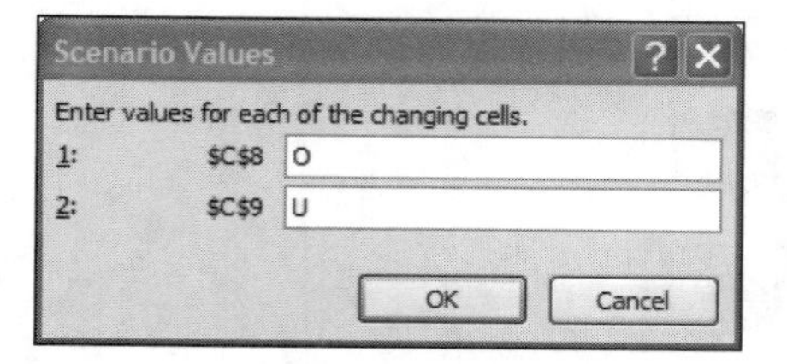

FIGURE C-16 Entering Opt-Up scenario input values

Click OK, which takes you back to the Scenario Manager window. Enter the other three Opt-Down, Pess-Up, and Pess-Down scenarios and related input values. When you finish, you should see that the names and changing cells for the four scenarios have been entered, as in Figure C-17.

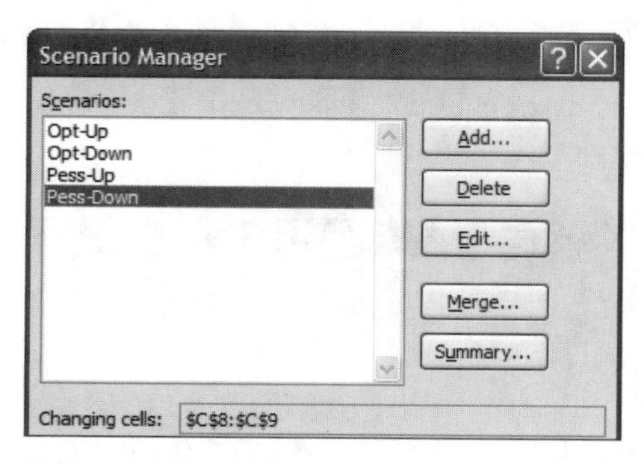

FIGURE C-17 Scenario Manager window with all scenarios entered

You can now create a summary sheet that shows the results of running the four scenarios. Click the Summary button to open the Scenario Summary window. You must provide Excel with the output cell addresses—they will be the same for all four scenarios. (The output *values* in those output cells change as input values are changed, but the addresses of the output cells do not change.)

Assume that you are interested primarily in the results that have accrued at the end of the two-year period. These results are your two Year 3 Summary of Key Results section cells for Net Income after Taxes and End-of-year cash on hand (D12 and D13). Type those addresses in the window's input area, as shown in Figure C-18. (Note that if the Result cells are noncontiguous, the address ranges can be entered and separated by a comma.)

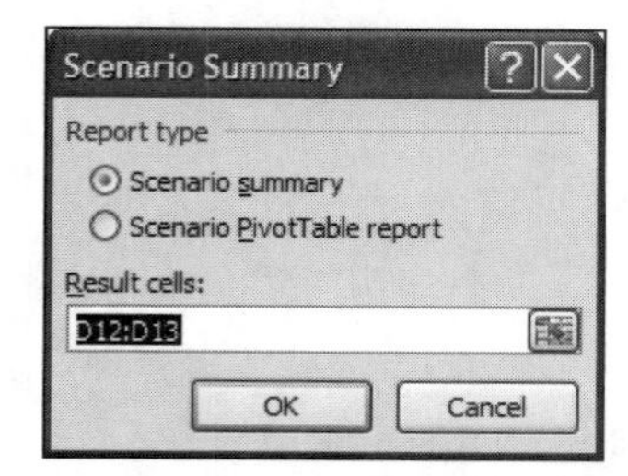

FIGURE C-18 Entering Result cells addresses in Scenario Summary window

Then click OK. Excel runs each set of inputs and collects results as it goes. (You do not see this on the screen.) Excel creates a new sheet called the Scenario Summary (denoted by the sheet's lower tab) and takes you there, as shown in Figure C-19.

	A	B	C	D	E	F	G	H
1								
2		Scenario Summary						
3				Current Values:	Opt-Up	Opt-Down	Pess-Up	Pess-Down
5		Changing Cells:						
6			$C$8	O	O	O	P	P
7			$C$9	U	U	D	U	D
8		Result Cells:						
9			$D$12	$728,909	$728,909	$1,085,431	$441,964	$752,953
10			$D$13	$1,535,753	$1,535,753	$2,045,679	$1,098,681	$1,552,944
11		Notes: Current Values column represents values of changing cells at						
12		time Scenario Summary Report was created. Changing cells for each						
13		scenario are highlighted in gray.						

FIGURE C-19 Scenario Summary sheet created by Scenario Manager

One slightly annoying visual element is that the Current Values in the spreadsheet are given an output column that duplicates one of the four defined scenarios. To delete the extra column, select the Home tab, select the Delete icon's drop-down arrow within the Cells group, and select Delete Sheet Columns.

NOTE

To delete a sheet's row, follow the steps in the preceding sentence, but select Delete Sheet Rows instead of Delete Sheet Columns.

Another annoyance is that column A goes unused. You can click and delete it as you've been doing to move everything to the left. That should make columns of data easier to see on the screen without scrolling. Other ways to make the worksheet easier to read include: (1) entering words in column A to describe the input and output cells, (2) centering cell values by using the Center Text icon in the Home tab's Alignment group, and (3) showing data in Currency format by using the Number Format drop-down menu within the Home tab's Number group.

When you finish, your summary sheet should resemble the one shown in Figure C-20.

	A	B	C	D	E	F
1	Scenario Summary					
2			Opt-Up	Opt-Down	Pess-Up	Pess-Down
4	Changing Cells:					
5	Economic Outlook	$C$8	O	O	P	P
6	Purchase Price Outlook	$C$9	U	D	U	D
7	Result Cells:					
8	Net income after taxes	$D$12	$728,909	$1,085,431	$441,964	$752,953
9	End-of-year cash on hand	$D$13	$1,535,753	$2,045,679	$1,098,681	$1,552,944

FIGURE C-20 Scenario Summary sheet after formatting

Note that column C shows the Optimistic-Up case. The Net Income after taxes in that scenario is $728,909, and End-of-year cash on hand is $1,535,753. Columns D, E, and F show the other scenario results.

As an important postscript to this exercise, note that DSS spreadsheets are used to guide decision making, which means that the spreadsheet's results must be interpreted in some way. Here are two practice questions based on the results in Figure C-20: With that data, what combination of Year 3 Net Income after taxes and End-of-year cash on hand would be best? Clearly, Optimistic-Down (O-D) is the best result, right? It yields the highest income and highest cash. What is the worst combination? Pessimistic-Up (P-U), right? It yields the lowest income and lowest cash.

Results are not always that easy to interpret, but the analytical method is the same. You have a complex situation that you cannot understand very well without software assistance. You build a model of the situation in the spreadsheet, enter the inputs, collect the results, and then interpret the results to help with decision making.

Summary Sheets

When you perform Scenario Manager spreadsheet case studies, you'll need to manipulate summary sheets and their data. Next, you will look at some of those operations.

Rerunning the Scenario Manager

The Scenario Summary sheet does not update itself when the spreadsheet formulas or inputs change. To see an updated Scenario Summary sheet, you must rerun the Scenario Manager. To rerun the Scenario Manager, click the Summary button in the Scenario Manager dialog box, and then click OK. Another summary sheet is created; it does not overwrite a prior one.

Deleting Unwanted Scenario Manager Summary Sheets

Suppose you want to delete a summary sheet. With the summary sheet on the screen, follow this procedure: (1) Select the Home tab, (2) within the Cells group, select the Delete icon's drop-down arrow, and (3) select Delete Sheet. When asked if you really want to delete this sheet, click Delete.

Charting Summary Sheet Data

The summary sheet results can be conveniently charted using the Chart Wizard, as discussed in Tutorial E.

Copying Summary Sheet Data to the Clipboard

If you want to put the summary sheet data into the Clipboard to use in a Microsoft Word document, follow these steps:

1. Select the data range.
2. Copy the data range into the Clipboard by following the copying operation described earlier in this tutorial.
3. Open your Microsoft Word document.
4. Click the cursor where you want the upper-left part of the data to be positioned.
5. Paste the data into the document by selecting Paste in the Home tab's Clipboard group.

PRACTICE USING SCENARIO MANAGER

Suppose you have an uncle who works for a large company. He has a good job and makes a decent salary (currently, $80,000 a year). At age 65, which will be in 2017, he can retire from his company and start drawing his pension.

However, the company has an early-out plan in which employees are asked to quit (called "preretirement"). The company then pays those employees a bonus in the year they quit and each year thereafter, up to the official retirement date, which is through the year 2016 for your uncle. Then employees start to receive their actual pension—in your uncle's case, in 2017. This early-out program would let your uncle leave the company before 2017. Until then, he could find a part-time hourly job to make ends meet and then leave the workforce entirely in 2017.

The opportunity to leave early is open through 2016. That means your uncle could stay with the company in 2011, then leave the company any year in the period 2012 to 2016 and get the early-out bonuses in the years he is retired. If he retires in 2012, he would lose the 2011 bonus. If he retires in 2013, he would lose the 2011 and 2012 bonuses. This would continue all the way through 2016.

Another factor in your uncle's thinking is whether to continue his country club membership. He likes the club, but it is a cash drain. The early-out decision can be assessed each year, but the decision about the country club membership must be made now; if your uncle does not withdraw in 2011, he says he will remain a member (and incur costs) through 2016.

Your uncle has called you in to make a spreadsheet model of his situation in the Scenario Manager. Your spreadsheet would let him play "what if" with the preretirement and country club possibilities and see his various projected personal finance results for 2011–2016. With each scenario, your uncle wants to know what "cash on hand" will be available for each year in the period.

Complete the spreadsheet for your uncle. Your Summary of Key Results, Calculations, and Income and Cash Flow Statements section cells must show values *by cell formula*. In other words, do not hard-code amounts in those sections. Also do not use the address of a cell if its contents are NA in any of your formulas. Set up your spreadsheet skeleton as shown in the figures that follow. Name your spreadsheet **UNCLE.xlsx**.

Constants Section

Your spreadsheet should have the constants shown in Figure C-21. An explanation of line items follows the figure.

	A	B	C	D	E	F	G	H
1	**Your Uncle's Early Retirement Decision**							
2	**Constants**	**2010**	**2011**	**2012**	**2013**	**2014**	**2015**	**2016**
3	Salary increase factor		3%	3%	2%	2%	1%	1%
4	Part time wages expected		$10,000	$10,200	$10,500	$10,800	$11,400	$12,000
5	Buy out amount		$30,000	$25,000	$20,000	$15,000	$5,000	$0
6	Cost of living (not retired)		$41,000	$42,000	$43,000	$44,000	$45,000	$46,000
7	Country club dues		$12,000	$13,000	$14,000	$15,000	$16,000	$17,000

FIGURE C-21 Constants section values

- Salary increase factor—Your uncle's salary at the end of 2010 will be $80,000. As you can see, raises are expected each year; for example, a 3% raise is expected in 2011. If your uncle does not retire, he will get his salary and a small raise for the year.
- Part time wages expected—Your uncle has estimated his part-time wages as if he were retired and working part-time from 2011 to 2016.
- Buy out amount—The amounts for the company's preretirement buyout plan are shown. If your uncle retires in 2011, he gets $30,000, $25,000, $20,000, $15,000, $5,000, and zero in the years 2011 to 2016, respectively. If he leaves in 2012, he will give up the $30,000 payment for 2011, but will get $25,000, $20,000, $15,000, $5,000, and zero in the years 2012 to 2016, respectively.
- Cost of living (not retired)—Your uncle has estimated how much cash he needs to meet his living expenses, assuming he continues to work for the company. His cost of living would be $41,000 in 2011, increasing each year thereafter.
- Country club dues—Country club dues are $12,000 for 2011. They increase each year thereafter.

Inputs Section

Your spreadsheet should have the inputs shown in Figure C-22. An explanation of line items follows the figure.

	A	B	C	D	E	F	G	H
9	**Inputs**	**2010**	**2011**	**2012**	**2013**	**2014**	**2015**	**2016**
10	Retired [R] or Working [W]							
11	Stay in club? [Y] or [N]			NA	NA	NA	NA	NA

FIGURE C-22 Inputs section

- Retired or Working—Enter an R if your uncle retires in a year or a W if he is still working. If he is working through 2016, you should enter the pattern WWWWWW. If his retirement is in 2011, you should enter the pattern RRRRRR. If he works for three years and then retires in 2014, you should enter the pattern WWWRRR.
- Stay in club?—If your uncle stays in the club from 2011 to 2016, you should enter a Y. If your uncle leaves the club in 2011, you should enter an N. The decision applies to all years.

Summary of Key Results Section

Your spreadsheet should show the results in Figure C-23.

	A	B	C	D	E	F	G	H
13	**Summary of Key Results**	**2010**	**2011**	**2012**	**2013**	**2014**	**2015**	**2016**
14	End-of-year cash on hand	NA						

FIGURE C-23 Summary of Key Results section

Each year's End-of-year cash on hand value is echoed from cells in the spreadsheet body.

Calculations Section

Your spreadsheet should calculate, by formula, the values shown in Figure C-24. Calculated amounts are used later in the spreadsheet. An explanation of line items follows the figure.

	A	B	C	D	E	F	G	H
16	**Calculations**	**2010**	**2011**	**2012**	**2013**	**2014**	**2015**	**2016**
17	Tax rate							
18	Cost of living							
19	Yearly salary or wages	$80,000						
20	Country club dues paid							

FIGURE C-24 Calculations section

- Tax rate—Your uncle's tax rate depends on whether he is retired. Retired people have lower overall tax rates. If he is retired in a year, your uncle's rate is expected to be 15 percent of income before taxes. In a year in which he works full-time, the rate will be 30 percent.
- Cost of living—In any year that your uncle continues to work for the company, his cost of living is the amount shown in the Cost of living (not retired) field in the Constants section in Figure C-21. But, if he chooses to retire, his cost of living is $15,000 less than the amount shown in the figure.
- Yearly salary or wages—If your uncle keeps working, his salary increases each year. The year-to-year percentage increases are shown in the Constants section. Thus, salary earned in 2011 would be more than that earned in 2010, salary earned in 2012 would be more than that earned in 2011, and so on. If your uncle retires in a certain year, he will make the part-time wages shown in the Constants section.
- Country club dues paid—If your uncle leaves the club, the dues are zero each year; otherwise, the dues are as shown in the Constants section.

The Income and Cash Flow Statements Section

This section begins with the cash on hand at the beginning of the year, followed by the income statement, and concluding with the calculation of cash on hand at the end of the year. The format is shown in Figure C-25. An explanation of line items follows the figure.

	A	B	C	D	E	F	G	H
22	**Income and Cash Flow Statements**	**2010**	**2011**	**2012**	**2013**	**2014**	**2015**	**2016**
23	Beginning-of-year cash on hand							
24	Salary or wages							
25	Buy out income							
26	Total Cash Inflow							
27	Country club dues paid							
28	Cost of living							
29	Total Costs							
30	Income before taxes							
31	Income tax expense							
32	Net income after taxes							
33	End-of-year cash on hand	$30,000						

FIGURE C-25 Income and Cash Flow Statements section

- Beginning-of-year cash on hand—The End-of-year cash on hand at the end of the prior year.
- Salary or wages—A yearly calculation, which can be echoed here.
- Buy out income—The year's buyout amount if your uncle is retired that year.

- Total Cash inflow—The sum of salary or part-time wages and buyout amounts.
- Country club dues paid—A calculated amount, which can be echoed here.
- Cost of living—A calculated amount, which can be echoed here.
- Total Costs—These outflows are the sum of the Cost of living and Country club dues paid.
- Income before taxes—This amount is the Total Cash inflow, less Total Costs (outflows).
- Income tax expense—This amount is zero when Income before taxes is zero or less; otherwise, the calculated tax rate is applied to the Income before taxes.
- Net income after taxes—Income before taxes, less tax expense.
- End-of-year cash on hand—The Beginning-of-year cash on hand plus the year's Net income after taxes.

Scenario Manager Analysis

Set up the Scenario Manager and create a Scenario Summary sheet. Your uncle wants to look at the following four possibilities:

- Retire in 2011, staying in the club ("Loaf-In")
- Retire in 2011, leaving the club ("Loaf-Out")
- Work three more years and retire in 2014, staying in the club ("Delay-In")
- Work three more years and retire in 2014, leaving the club ("Delay-Out")

You can enter noncontiguous cell ranges as follows: C20:F20, C21, C22 (cell addresses are examples). The output cell should be only the 2016 End-of-year cash on hand cell.

Your uncle will choose the option that yields the highest 2016 End-of-year cash on hand. You must look at your Scenario Summary sheet to see which strategy yields the highest amount.

To check your work, you should attain the values shown in Figure C-26. (You can use the labels that Excel provides in the far left column or change the labels, as shown in Figure C-26.)

	A	B	C	D	E	F
1	Scenario Summary					
2			Loaf-In	Loaf-Out	Delay-In	Delay-Out
4	Changing Cells:					
5	Retire or Work, 2011	$C$10	R	R	W	W
6	Retire or Work, 2012	$D$10	R	R	W	W
7	Retire or Work, 2013	$E$10	R	R	W	W
8	Retire or Work, 2014	$F$10	R	R	R	R
9	Retire or Work, 2015	$G$10	R	R	R	R
10	Retire or Work, 2016	$H$10	R	R	R	R
11	In club? 2011-2016	$C$11	Y	N	Y	N
12	Result Cells (2016):					
13	End-of-year cash on hand, 2016	$H$14	-$68,400	$15,195	$8,389	$83,689

FIGURE C-26 Scenario Summary

REVIEW OF EXCEL BASICS

In this section, you'll begin by reviewing how to perform some basic operations. Then you'll work through more cash flow calculations. Working through this section will help you complete the spreadsheet cases in this book.

Basic Operations

To begin, you'll review the following topics: formatting cells, showing Excel cell formulas, understanding circular references, using the AND and the OR functions in IF statements, and using nested IF statements.

Formatting Cells

You may have noticed that some data in this tutorial's first spreadsheet was centered in the cells. Follow these steps to center data in cells:

1. Highlight the cell range to format.
2. Select the Home tab.

3. In the Alignment group, select the Middle Align icon to change the vertical alignment.
4. In the Alignment group, select the Center icon to change the horizontal alignment.

You can also put a border around cells, which might be desirable for highlighting Inputs section cells. Follow these steps:

1. Highlight the cell that needs a border.
2. Select the Home tab.
3. In the Font group, select the drop-down arrow of the Bottom Border icon.
4. Choose the desired border's menu choice; All Borders would be best for your purposes.

You can format numerical values for Currency by following these steps:

1. Highlight the cell or range of cells that should be formatted.
2. Select the Home tab.
3. In the Number group, use the Number Format drop-down arrow to select Currency.

You can format numerical values for decimal places using this procedure:

1. Highlight the cell or range of cells that should be formatted.
2. Select the Home tab.
3. In the Number group, click the Increase Decimal icon once to add one decimal value. Click the Decrease Decimal icon to eliminate a decimal value.

You can use the available automatic styles to change the appearance of the text:

1. Highlight the cell or range of cells that you want to change.
2. Select the Home tab.
3. Within the Styles group, click the More button, if necessary, then choose the style you want to apply, such as Heading 1 or Title in the Heading and Titles group.

You can increase or decrease the indent used for your row headings, which makes the text under each section much easier to see:

1. Highlight the cell or range of cells that you want to change.
2. Select the Home tab.
3. In the Alignment group, click the Increase Indent button.

Showing Excel Cell Formulas

If you want to see Excel cell formulas, follow these steps:

1. Press the Ctrl key and the back quote key (') at the same time. The back quote faces the opposite direction from a normal quotation mark; on most keyboards, it shares the key with the tilde (~) mark.
2. To restore, press the Ctrl and back quote keys again.

Understanding a Circular Reference

A formula has a circular reference when the reference *refers to itself either directly or indirectly*. Excel cannot properly evaluate such a formula. The problem is best described by an example. Suppose the formula in cell C18 is =C18–C17. Excel is trying to compute a value for cell C18, so it must evaluate the formula, then put the result on the screen in C18. Excel tries to subtract the contents of C17 from the contents of C18, but nothing is in C18. Can you see the circularity? To establish a value for C18, Excel must know what is in C18. The process is circular—hence, the term *circular reference*. In the example, the formula in C18 refers to C18. As another simple example, consider a formula in one cell that refers to a formula in a second cell, and the formula in the second cell that refers to the formula in the first cell. For example, cell C7 is =C6, and cell C6 is =C7. Excel points out circular references, if any exist, when you choose Open for a spreadsheet. Excel also points out circular references as you insert them during the building of a spreadsheet. The software alerts you to circular references by opening at least one Help window and by drawing arrows between cells involved in the offending formula. You can close the windows, but that will not fix the situation. You *must* fix the formula that has the circular reference if you want the spreadsheet to give you accurate results.

Using the AND Function and the OR Function in IF Statements

An IF statement has the following syntax:

IF(test condition, result if test is True, result if test is False)

The test conditions in this tutorial's IF statements tested only one cell's value, but a test condition can test more than one value of a cell.

Here is an example from this tutorial's first spreadsheet, in which selling price was a function of the economy. Assume for the sake of illustration that Year 2's selling price per unit depends on the economy *and* on the purchase-price outlook. There are two possibilities: (1) If the economic outlook is optimistic *and* the company's purchase-price outlook is down, the selling price will be 110% times the prior year's price. (2) In all other cases, the selling price will be 103% times the prior year's price. The first possibility's test requires two things to be true *at the same time*: C8 = "O" *AND* C9 = "D." To implement the test, the AND() function is needed. The code in cell C17 would be as follows:

=IF(AND($C$8 = "O", $C$9 = "D"), B17 * 110%, B17 * 103%)

When the test that uses the AND() function evaluates to True, the result is B17 * 110%. When the test evaluates to False, the result is the second possibility's outcome: B17 * 103%.

Now suppose the first possibility is as follows: If the economic outlook is optimistic *or* the purchase-price outlook is down, the selling price will be 110% times the prior year's price. Assume in all other cases that the selling price will be 103% times the prior year's price. Now the test requires *only one of* two things to be true: C8 = "O" *OR* C9 = "D." To implement that test, the OR() function is needed. The code in cell C17 would be:

=IF(OR($C$8 = "O", $C$9 = "D"), B17 * 110%, B17 * 103%)

Using IF Statements Inside IF Statements

Recall from the previous section that an IF statement has this syntax:

IF(test condition, result if test is True, result if test is False)

In the examples shown thus far, only two courses of action were possible, so only one test was needed in the IF statement. However, there can be more than two courses of action; if so, the "result if test is False" clause needs to show further testing. Look at the following example.

Assume again that the Year 2 selling price per unit depends on the economy and the purchase-price outlook. Here is the logic: (1) If the economic outlook is optimistic *and* the purchase-price outlook is down, the selling price will be 110% times the prior year's price. (2) If the economic outlook is optimistic *and* the purchase-price outlook is up, the selling price will be 107% times the prior year's price. (3) In all other cases, the selling price will be 103% times the prior year's price. The code in cell C17 would be as follows:

=IF(AND($C$8 = "O", $C$9 = "D"), B17 * 110%,
IF(AND($C$8 = "O", $C$9 = "U"), B17 * 107%, B17 * 103%))

The first IF statement tests to see if the economic outlook is optimistic and the purchase-price outlook is down. If not, further testing is needed to see whether the economic outlook is optimistic and the purchase-price outlook is up, or whether some other situation prevails.

NOTE

The line is broken in the previous example because the page is not wide enough, but in Excel, the formula would appear on one line. The embedded "IF" is not preceded by an equal sign.

Cash Flow Calculations: Borrowings and Repayments

The Scenario Manager cases in this book require you to account for money that the company borrows or repays. Borrowing and repayment calculations are discussed next. At times, you will be asked to think about a question and fill in the answer. Correct responses are found at the end of this section.

To work through the Scenario Manager cases that follow, you must assume two things about a company's borrowing and repayment of debt. First, assume that the company wants to have a certain minimum cash level at the end of a year (and thus at the start of the next year). Second, assume that a bank will provide a loan to make up the shortfall if year-end cash falls short of the desired minimum cash level.

Here are some examples to test your understanding. Assume that NCP stands for "net cash position" and equals beginning-of-year cash plus net income after taxes for the year. In other words, the NCP is the cash available at year end, before any borrowing or repayment. For the three examples in Figure C-27, compute the amounts the company needs to borrow to reach its minimum year-end cash level.

Example	NCP	Minimum Cash Required	Amount to Borrow
1	$50,000	$10,000	?
2	$8,000	$10,000	?
3	-$20,000	$10,000	?

FIGURE C-27 Examples of borrowing

One additional assumption you can make is that the company will use its excess cash at year end to pay off as much debt as possible without going below the minimum-cash threshold. Excess cash is the NCP *less* the minimum cash required on hand—amounts over the minimum are available to repay any debt.

In the examples shown in Figure C-28, compute excess cash and then compute the amount to repay. To aid your understanding, you also may want to compute ending cash after repayments.

Example	NCP	Minimum Cash Required	Beginning-of-Year Debt	Repay	Ending Cash
1	$12,000	$10,000	$4,000	?	?
2	$12,000	$10,000	$10,000	?	?
3	$20,000	$10,000	$10,000	?	?
4	$20,000	$10,000	$0	?	?
5	$60,000	$10,000	$40,000	?	?
6	-$20,000	$10,000	$10,000	?	?

FIGURE C-28 Examples of repayment

In this section's Scenario Manager cases, your spreadsheet will need two bank financing sections beneath the Income and Cash Flow Statements section: The first section will calculate any needed borrowing or repayment at year's end to compute year-end cash. The second section will calculate the amount of debt owed at the end of the year, after borrowing or repayment of debt.

The first new section, in effect, extends the end-of-year cash calculation, which was shown in Figure C-13. Previously, the amount equaled cash at the beginning of the year plus the year's net income. Now the calculation will include cash obtained by borrowing and cash repaid. Figure C-29 shows the structure of the calculation.

	A	B	C	D
30	Net Cash Position (NCP) Beginning-of-year cash on hand plus Net income after taxes	NA		
31	Borrowing from bank	NA		
32	Repayment to bank	NA		
33	End-of-year cash on hand	$10,000		

FIGURE C-29 Calculation of end-of-year cash on hand

The heading in cell A30 was previously End-of-year cash on hand in Figure C-13, but Borrowing increases cash and Repayment of debt decreases cash. So, End-of-year cash on hand is now computed two rows down (in C33 for Year 2 in the example). The value in row 30 must be a subtotal for the Beginning-of-year cash on hand plus the year's Net income after taxes. That subtotal is called the Net Cash Position (NCP). (Note that the formula in cell C22 for Beginning-of-year cash on hand was =B28, but now it is =B33. It is copied to the right, as before, for the next year.)

The second new section computes End-of-year debt and is called Debt Owed, as shown in Figure C-30.

	A	B	C	D
35	**Debt Owed**	**Year 1**	**Year 2**	**Year 3**
36	Beginning-of-year debt owed	NA		
37	Borrowing from bank	NA		
38	Repayment to bank	NA		
39	End-of-year debt owed	$15,000		

FIGURE C-30 Debt Owed section

As you can see in Figure C-30, $15,000 was owed at the end of Year 1. The End-of-year debt owed equals the Beginning-of-year debt owed plus any new Borrowing from bank (which increases debt owed), less any Repayment to bank (which reduces it). So, in the example, the formula in cell C39 would be:

=C36+C37−C38

Assume that the amounts for Borrowing from bank and Repayment to bank are calculated in the first new section. Thus, the formula in cell C37 would be =C31. The formula in cell C38 would be =C32. (Beginning-of-year debt owed is equal to the debt owed at the end of the prior year, of course. The formula in cell C36 for Beginning-of-year debt owed would be an echoed formula. Can you see what it would be? It's an exercise for you to complete.

Now that you have seen how the borrowing and repayment data is shown, we can discuss the logic of the borrowing and repayment formulas.

Calculation of Borrowing from Bank

The logic of this in English is:

If (cash on hand before financing transactions is greater than the minimum cash required,
then borrowing is not needed;
else, borrow enough to get to the minimum).

Or (a little more precisely):

If (NCP is greater than the minimum cash required,
then Borrowing from bank = 0;
else, borrow enough to get to the minimum).

Suppose the desired minimum cash at year end is $10,000, which is a constant in your spreadsheet's cell C6. Assume that the NCP is shown in your spreadsheet's cell C30. The formula (getting closer to Excel) would be as follows:

IF(NCP > Minimum Cash, 0; otherwise, borrow enough to get to the minimum).

You have cell addresses that stand for NCP (cell C30) and Minimum Cash (C6). To develop the formula for cell C31, substitute the cell addresses for NCP and Minimum Cash. The harder logic is for the "else" clause. At this point, you should look ahead to the Borrowing answers in Figure C-31. In Example 2, $2,000 was borrowed. Which cell was subtracted from which other cell to calculate that amount? Substitute cell addresses in the Excel formula for Year 2's borrowing formula in cell C31:

=IF(>=   , 0,   −   )

The answer is at the end of this section in Figure C-33.

Calculation of Repayment to Bank

The logic of this in English is:

IF(beginning of year debt = 0, repay 0 because nothing is owed, but
 IF(NCP is less than the min, repay zero, because you must *borrow*, but
  IF(extra cash equals or exceeds the debt, repay the whole debt,
   ELSE (to stay above the min, repay only the extra cash))))

Look at the following formula. Assume that the repayment will be in cell C32. Assume also that debt owed at the beginning of the year is in cell C36 and minimum cash is in cell C6. Substitute cell addresses for concepts to complete the formula for Year 2 repayment. (Clauses appear on different lines because of page width limitations.)

=IF(= 0, 0,
 IF(<=, 0,
  IF((–) >=,
  (–))))

The answer is shown at the end of this section in Figure C-34.

Answers to Questions about Borrowing and Repayment Calculations

Figure C-31 and Figure C-32 answer the questions about borrowing and repayment calculations.

Example	NCP	Minimum Cash Required	Amount to Borrow	Comments
1	$50,000	$10,000	$0	NCP > Min
2	$8,000	$10,000	$2,000	Need $2,000 to get to Min (10,000 - 8,000)
3	-$20,000	$10,000	$30,000	Need $30,000 to get to Min (10,000 - (-20,000))

FIGURE C-31 Answers to examples of borrowing

Example	NCP	Minimum Cash Required	Beginning-of-Year Debt	Repay	Ending Cash
1	$12,000	$10,000	$4,000	$2,000	$10,000
2	$12,000	$10,000	$10,000	$2,000	$10,000
3	$20,000	$10,000	$10,000	$10,000	$10,000
4	$20,000	$10,000	$0	$0	$20,000
5	$60,000	$10,000	$40,000	$40,000	$20,000
6	-$20,000	$10,000	$10,000	$0	NA

FIGURE C-32 Answers to examples of repayment

Note the following points about the repayment calculations shown in Figure C-32.

- In Examples 1 and 2, only $2,000 is available for debt repayment (12,000 – 10,000) to avoid going below the minimum cash.
- In Example 3, cash available for repayment is $10,000 (20,000 – 10,000), so all beginning debt can be repaid, leaving the minimum cash.
- In Example 4, no debt is owed, so no debt need be repaid.
- In Example 5, cash available for repayment is $50,000 (60,000 – 10,000), so all beginning debt can be repaid, leaving more than the minimum cash.
- In Example 6, no cash is available for repayment. The company must borrow.

Figure C-33 and Figure C-34 show the calculations for borrowing and repayment of debt.

```
=IF(C30 >= C6, 0 , C6 - C30)
```

FIGURE C-33 Calculation of borrowing

```
=IF( C36 = 0, 0, IF( C30 <= C6, 0 , IF ( C30 - C6 >= C36, C36, C30 - C6)))
```

FIGURE C-34 Calculation of repayment

Saving Files after Using Microsoft Excel

As you work, you should make a habit of saving your files periodically. The top of the screen should have a Quick Access toolbar, which includes the Save button by default. (The button looks like a diskette.) You can save your work by clicking the Save button or the Office button, which is the relatively large button in the upper-left corner of the screen. You can also choose the Save menu. The first time you save, you will use the Save As window to specify the following:

- A drive (using the My Documents drop-down menu)
- A filename (using the File name text box)
- A file type (by default, Excel 2007 creates an .xlsx file, but you can save in the Excel 2003 format or in other formats)

You will see the Save As window every time you choose Save As (not Save).

At the end of your Excel session, save your work using the following three steps:

1. Save the file one last time.
2. Click the Close button, which is the X in the upper-right corner of your window. If you are saving to a secondary disk or USB key, make sure it is still in its drive when you close the file. Closing the file takes the work off the screen, so do not remove the disk from its drive if the work is still on the screen. Otherwise, you may lose your work.
3. Using the Office button, choose Exit Excel to return to Windows.

CASE **6**

THE MADE4U DOUGHNUTS MUFFIN DECISION

Decision Support Using Excel

PREVIEW

Your doughnut shop is doing well, and you are thinking about adding muffins as a complementary product. You would have to borrow money to finance the equipment needed to make muffins. In this case, you will use Microsoft Excel to see if the added product line would make financial sense.

PREPARATION

- Review spreadsheet concepts discussed in class and in your textbook.
- Complete any exercises that your instructor assigns.
- Complete any part of Tutorial C that your instructor assigns. You may need to review the use of If statements and the section called "Cash Flow Calculations: Borrowings and Repayments."
- Review file-saving procedures for Windows programs.
- Refer to Tutorial E as necessary.

BACKGROUND

You own the Made4U doughnut shop, which sells delicious, fresh, hot doughnuts. Your shop is on Main Street in your hometown. You have plenty of traffic both from pedestrians and customers in cars. Sales are brisk, and your business is doing well.

Other shops on Main Street sell baked goods, but none of them sell doughnuts. You notice from observation and from reading the trade journals that more people are eating muffins. You want to know if adding muffins to your menu will increase your profits.

Muffin baking is a separate process from frying and would require a significant amount of investment capital. You would borrow the money from your bank and install the equipment by the end of 2010.

In this case, you will use Excel to see whether Made4U can achieve profits and manage the debt that would result from adding muffins to the product line.

Your DSS would include the following inputs:

1. Your decision to sell muffins or not
2. The anticipated growth rate of your business
3. The "pricing power" you anticipate in the future

You set your prices by adding an amount called the "margin" to the item's unit cost. For example, if the unit cost to make a doughnut is 20 cents, you would add a nickel margin, meaning that the doughnut's selling price is 25 cents. In effect, your profit margin is the nickel. If you have "pricing power," then you have the ability to add a high margin. If you do not have pricing power, you cannot be aggressive with the margin you add.

Your DSS model needs to account for the effects of the input values on costs, selling prices, and other variables. Your model will let you develop "what-if" scenarios with the inputs, see the results, and then decide what to do.

ASSIGNMENT 1: CREATING A SPREADSHEET FOR DECISION SUPPORT

In this assignment, you will produce a spreadsheet that models the business decision. Then, in Assignment 2, you will write a memorandum to the bank about your analysis and recommendations. In Assignment 3, you will prepare and give an oral presentation of your analysis and recommendations.

First, you will create the spreadsheet model of the proposal's financial profile. The model covers the three years from 2011 to 2013. This section helps you set up each of the following spreadsheet components before entering cell formulas:

- Constants
- Inputs
- Summary of Key Results
- Calculations
- Income and Cash Flow Statements
- Debt Owed

A discussion of each section follows. *The spreadsheet skeleton is available for you to use; you can choose to type in it or not.* To access the spreadsheet skeleton, go to your data files, select Case 6, and then select **Made4U.xlsx**.

Constants Section

Your spreadsheet should include the constants shown in Figure 6-1. An explanation of the line items follows the figure.

	A	B	C	D	E
1	**Made4U Doughnuts Muffin Decision**				
2	**Constants**	**2010**	**2011**	**2012**	**2013**
3	Tax rate	NA	30%	30%	30%
4	Minimum cash needed to start year	NA	$1,000,000	$1,000,000	$1,000,000
5	Fixed administrative expense	NA	$225,000	$235,000	$245,000
6	Interest rate for year	NA	5.50%	6.00%	6.00%
7	Business days per year	NA	364	365	364
8	Cost per doughnut	NA	$0.15	$0.17	$0.19
9	Cost per cup of coffee	NA	$1.65	$1.75	$1.85
10	Cost per muffin	NA	$1.25	$1.50	$1.75
11	Average salary per worker	NA	$25,000	$26,000	$27,000

FIGURE 6-1 Constants section

- Tax rate—The tax rate is applied to income before taxes. The rate is expected to stay constant each year.
- Minimum cash needed to start year—You want to have at least $1 million in cash at the beginning of each year. Your banker will lend you the amount you need at the end of a year in order to begin the new year with $1 million.
- Fixed administrative expense—Rent, maintenance, insurance, electricity, and so on, which are expected to increase each year.
- Interest rate for year—The interest rate your banker will charge for any borrowing. The banker says that interest rates are expected to rise as the economy recovers.
- Business days per year—Your shop is open every day except Christmas Day. Notice that 2012 is a leap year, so you will be open 365 days that year.
- Cost per doughnut—The cost of the raw ingredients needed to make a doughnut, including flour, sugar, salt, flavorings, oil, milk, raisins, and nuts. The cost to make doughnuts has increased each year, and you expect the trend to continue.
- Cost per cup of coffee—The cost of coffee, sugar, artificial sweetener, milk, and cream, all of which are needed to make a cup of coffee. You buy only fairly traded, organic coffee, sugar, and milk, so your costs are higher than average.

- Cost per muffin—The cost of the raw ingredients for each muffin. The same ingredients used to make muffins are used to make doughnuts. Muffins are larger than doughnuts, so you will use more of each ingredient to make muffins. You anticipate increased costs for the next three years.
- Average salary per worker—The average salary is expected to increase each year, as shown.

Inputs Section

Your spreadsheet should include the following inputs for the years 2011, 2012, and 2013, as shown in Figure 6-2.

	A	B	C	D	E
13	Inputs	2010	2011	2012	2013
14	Sell muffins? (Y=Yes, N=No)	NA		NA	NA
15	Growth rate (H=High, M=Medium, L=Low)	NA			
16	Pricing Power (Y/N)?	NA		NA	NA

FIGURE 6-2 Inputs section

- Sell muffins?—Will you add muffins as a product line? Enter a Y or an N. The choice applies to all three years.
- Growth rate—Enter the rate of sales growth expected for each year. This input will later be used to estimate doughnut sales in the three years.
- Pricing Power?—Do you anticipate having pricing power in the three-year period? In other words, can you set your selling prices aggressively? Enter a Y or an N. The choice applies to all three years.

Summary of Key Results Section

Your spreadsheet should include the results shown in Figure 6-3. An explanation of each item follows the figure.

	A	B	C	D	E
18	Summary of key results	2010	2011	2012	2013
19	Net income after taxes	NA			
20	End-of-year cash on hand	$1,000,000			
21	End-of-year debt owed	$0			

FIGURE 6-3 Summary of key results section

For each year, your spreadsheet should show net income after taxes, cash on hand at the end of the year, and bank debt owed at the end of the year. The cells should be formatted as currency with zero decimals. These values are computed elsewhere in the spreadsheet and should be echoed here.

Calculations Section

You should calculate intermediate results that will be used in the income and cash flow statements that follow. Calculations, as shown in Figure 6-4, may be based on year-end 2010 values. When called for, use absolute referencing properly. An explanation of each item in this section follows the figure.

	A	B	C	D	E
23	**Calculations**	**2010**	**2011**	**2012**	**2013**
24	Average number of doughnuts sold per day	1,250			
25	Average cups of coffee sold per day	1,000			
26	Average number of muffins sold per day	NA			
27	Margin per item	NA			
28	Number of workers	3			

FIGURE 6-4 Calculations section

- Average number of doughnuts sold per day—This number is a function of the growth rate from the Inputs section. If the expected growth rate is High in a year, the number sold in that year will be 10% more than the number sold in the prior year. If the expected growth rate is Medium, the number sold in a year will be 3% more than in the prior year. If the expected growth rate is Low, the number sold will be 3% *less* than in the prior year.
- Average cups of coffee sold per day—This number is a function of the growth rate from the Inputs section. If the expected growth rate is High in a year, the number sold in that year will be 8% more than the number sold in the prior year. If the expected growth rate is Medium, the number sold in a year will be 5% more than in the prior year. If the expected growth rate is Low, the number sold will be 3% more than in the prior year.
- Average number of muffins sold per day—This number is a function of the decision to sell muffins, which is shown in the Inputs section. If you decide not to sell muffins, then of course this number will be zero. If you decide to sell muffins, then 500 will be sold per day in 2011, the number sold will be 3% greater in 2012 than in 2011, and the number sold will be 3% greater in 2013 than in 2012.
- Margin per item—This number is a function of the pricing power, which is shown in the Inputs section. If you will have pricing power, the margin per item will be 45 cents. Otherwise, the margin per item will be 40 cents.
- Number of workers—This number is a function of expected growth, which is shown in the Inputs section. If the annual growth rate is expected to be High, one worker more will be employed than in the prior year. If the annual growth rate is expected to be Medium, the number of workers will remain the same as in the prior year. If the yearly growth rate is expected to be Low, one worker *less* will be employed than in the prior year. In no year, however, will only one worker be employed; you cannot run the shop with just one worker.

Income and Cash Flow Statements

The forecast for net income and cash flow starts with the cash on hand at the beginning of the year. This is followed by the income statement and concludes with the calculation of cash on hand at year's end. For readability, format cells in this section as currency with zero decimals. Your spreadsheets should look like those shown in Figures 6-5 and 6-6. A discussion of each item in the section follows each figure.

	A	B	C	D	E
30	**Income and Cash Flow Statements**	**2010**	**2011**	**2012**	**2013**
31	Beginning-of-year cash on hand	NA			
32	Revenue	NA	NA	NA	NA
33	Yearly doughnut revenue	NA			
34	Yearly coffee revenue	NA			
35	Yearly muffin revenue	NA			
36	Total revenue	NA			
37	Costs	NA	NA	NA	NA
38	Yearly doughnut costs	NA			
39	Yearly coffee costs	NA			
40	Yearly muffin costs	NA			
41	Salary costs	NA			
42	Fixed administrative costs	NA			
43	Total costs	NA			
44	Income before interest and taxes	NA			
45	Interest expense	NA			
46	Income before taxes	NA			
47	Income tax expense	NA			
48	Net income after tax	NA			

FIGURE 6-5 Income and Cash Flow Statements section

- Beginning-of-year cash on hand—The cash on hand at the end of the prior year.
- Yearly doughnut revenue—This amount is a function of the selling price, the number of doughnuts sold per day, and the number of business days. Recall that the selling price equals the unit cost plus the margin.
- Yearly coffee revenue—This amount is a function of the selling price, the number of cups of coffee sold per day, and the number of business days. Recall that the selling price equals the unit cost plus the margin.
- Yearly muffin revenue—This amount is a function of the selling price, the number of muffins sold per day, and the number of business days. Recall that the selling price equals the unit cost plus the margin.
- Total revenue—This amount is the sum of doughnut, coffee, and muffin revenue.
- Yearly doughnut costs—This amount is a function of the number of doughnuts sold, the unit cost of a doughnut, and the number of business days.
- Yearly coffee costs—This amount is a function of the number of cups sold, the unit cost of a cup of coffee, and the number of business days.
- Yearly muffin costs—This amount is a function of the number of muffins sold, the unit cost of a muffin, and the number of business days.
- Salary costs—This amount is a function of the number of workers in the year and the average salary per worker.
- Fixed administrative costs—This amount is a constant that can be echoed here.
- Total costs—The sum of doughnut, coffee, muffin, salary, and fixed costs in the year.
- Income before interest and taxes—The difference between total revenue and total costs.
- Interest expense—The product of the debt owed at the beginning of the year and the annual interest rate on debt.
- Income before taxes—The income before interest and taxes minus the interest expense.
- Income tax expense—This value is zero if the income before taxes is zero or negative. Otherwise, income tax expense is the product of the year's tax rate and the income before taxes.
- Net income after tax—The difference between the income before taxes and income tax expense.

Line items for the year-end cash calculation are discussed next. In Figure 6-6, column B represents 2010, column C is for 2011, and so on. Year 2010 values are NA except for End-of-year cash on hand, which is $1 million.

	A	B	C	D	E
50	Net cash position (NCP)	NA			
51	Borrowing from bank	NA			
52	Repayment to bank	NA			
53	End-of-year cash on hand	$1,000,000			

FIGURE 6-6 End-of-year cash on hand section

- Net cash position (NCP)—The NCP at the end of a year equals the cash at the beginning of the year plus the year's net income after taxes.
- Borrowing from bank—Assume that a bank will lend you enough money at the end of the year to reach the minimum cash needed to start the next year. If the NCP is less than this minimum, you must borrow enough to start the next year with the minimum. Borrowing increases the cash on hand, of course.
- Repayment to bank—If the NCP is more than the minimum cash needed at the end of a year and debt is owed, you must pay off as much debt as possible (but not take cash below the minimum cash required to start the next year). Repayments reduce cash on hand, of course.
- End-of-year cash on hand—The NCP plus any borrowing and minus any repayments.

Debt Owed Section

This section shows a calculation of debt owed to the bank at year's end, as shown in Figure 6-7. An explanation of each item follows the figure.

	A	B	C	D	E
55	**Debt Owed**	**2010**	**2011**	**2012**	**2013**
56	Beginning-of-year debt owed	NA			
57	Borrowing from bank	NA			
58	Repayment to bank	NA			
59	End-of-year debt owed				

FIGURE 6-7 Debt Owed section

- Beginning-of-year debt owed—Debt owed at the beginning of a year equals the debt owed at the end of the prior year.
- Borrowing from bank—This amount has been calculated elsewhere and can be echoed to this section. Borrowing increases the amount of debt owed.
- Repayment to bank—This amount has been calculated elsewhere and can be echoed to this section. Repayments reduce the amount of debt owed.
- End-of-year debt owed—In 2011 through 2013, this is the amount owed at the beginning of a year, plus borrowing during the year, and minus repayments during the year. In 2010, this debt equals $500,000 if muffins will be added and zero if muffins will not be added. (Recall that adding muffins would require a bank-financed loan at the end of 2010.)

ASSIGNMENT 2: USING THE SPREADSHEET FOR DECISION SUPPORT

You complete the case by using the spreadsheet to gather data needed to determine your strategy and by documenting recommendations in a memorandum.

You think that a low growth rate is unlikely throughout the three years. You think that growth will either be medium in all years or will start low and then improve. You think that the following six scenarios will help answer your questions:

1. Muffins-Medium-No Power—You add muffins and enjoy medium growth in all years, but you have no pricing power.

2. Muffins-Increasing-No Power—You add muffins and see low growth in 2011, medium growth in 2012, and high growth in 2013. However, you have no pricing power.
3. Muffins-Medium-Power—You add muffins, enjoy medium growth in all years, and have pricing power.
4. Muffins-Increasing-Power—You add muffins and see low growth in 2011, medium growth in 2012, and high growth in 2013. You also have pricing power.
5. No Muffins-Medium-Power—You do not add muffins, and you enjoy medium growth and pricing power in all years.
6. No Muffins-Increasing-Power—You do not add muffins, and you see low growth in 2011, medium growth in 2012, and high growth in 2013. You also have pricing power.

The primary business question is whether muffins should be added as a product. You will use your spreadsheet to gather data on this issue and the following related questions. The data will help you decide whether to add muffins.

1. Are net income and debt significantly different, with and without muffins?
2. If muffins are added, how quickly will the $500,000 loan be paid off? Assume that you think you can pay off the loan in six years. You can estimate this by seeing how much of the loan remains after three years. In which scenarios does it appear that the loan will be paid off within six years?
3. How important is it to have pricing power? Compare the results of two similar scenarios. For example, how different are the results between scenarios 1 and 3 or between scenarios 2 and 4?

Assignment 2A: Using the Spreadsheet to Gather Data

You have built the spreadsheet to model the business situation. For each of the six scenarios listed earlier, you want to know the net income after taxes, the end-of-year cash on hand, and the end-of-year debt owed in 2013.

You will run "what-if" scenarios with the six sets of input values using the Scenario Manager. (See Tutorial C for details on using the Scenario Manager.) Set up the six scenarios. Your instructor may ask you to use conditional formatting to make sure your input values are proper. (Note that in the Scenario Manager you can enter noncontiguous cell ranges, such as C19, D19, C20:F20.)

The three output cells are the 2013 net income after taxes, end-of-year cash on hand, and end-of-year debt owed from the Summary of Key Results section. Run the Scenario Manager to gather the data in a report. When you finish, print the spreadsheet with the input for any of the scenarios, print the Scenario Manager summary sheet, and then save the spreadsheet file for the last time.

Assignment 2B: Documenting Your Recommendations in a Memorandum

Use Microsoft Word to write a brief memorandum to your banker. State the results of your analysis and give your recommendations about adding muffins. (Recall that adding muffins requires a $500,000 loan.) Observe the following requirements:

- Set up your memorandum as described in Tutorial E.
- In the first paragraph, briefly define your business and state the purpose of your analysis.
- Next, describe your results and state your recommendations, which should include answers to your questions about adding muffins.
- Support your statements graphically, as your instructor requires. If you used the Scenario Manager, your instructor may ask you to return to Excel and copy the Scenario Manager summary sheet results into the memorandum. (See Tutorial C for details on this procedure.) Your instructor might also ask you to make a summary table in Word based on the Scenario Manager summary sheet results. (This procedure is described in Tutorial E.)

Your table should resemble the format shown in Figure 6-8.

Scenario	2013 Net Income	2013 Cash on Hand	2013 Debt Owed
Muffins-Medium-No Power			
Muffins-Increasing-No Power			
Muffins-Medium-Power			
Muffins-Increasing-Power			
No Muffins-Medium-Power			
No Muffins-Increasing-Power			

FIGURE 6-8 Format of table to insert in memorandum

ASSIGNMENT 3: GIVING AN ORAL PRESENTATION

Your instructor may ask you to present your analysis and recommendations in an oral presentation. If so, assume that your banker wants you to explain your analysis and recommendations to her in 10 minutes or less. Use visual aids or handouts that you think are appropriate. See Tutorial F for tips on preparing and giving an oral presentation.

DELIVERABLES

Assemble the following deliverables for your instructor:

1. Printout of your memorandum
2. Spreadsheet printouts
3. Electronic media such as a USB key or CD that contains your Word memo and Excel spreadsheet file

Staple the printouts together with the memorandum on top. If you have more than one .xlsx file on your electronic media, write your instructor a note that identifies your model's .xlsx file.

CASE **7**

THE U.S. BUDGET PROJECTION

Decision Support Using Excel

PREVIEW

The U.S. government has been spending more money than it receives in taxes. The deficit is funded by loans. Future U.S. policy will depend on how big the deficit becomes. In this case, you will use Excel to create a projection of the federal deficit through 2020.

PREPARATION

- Review spreadsheet concepts discussed in class and in your textbook.
- Complete any exercises that your instructor assigns.
- Complete any part of Tutorial C that your instructor assigns.
- Review file-saving procedures for Windows programs.
- Refer to Tutorial E as necessary.

BACKGROUND

The U.S. government has been running a sizeable deficit in recent years. That is, the United States has been spending more than it receives in taxes. The difference (called the "deficit") is funded by loans. The U.S. borrows by issuing interest-bearing bonds to wealthy people and to foreign governments with cash to invest.

At the end of 2010, the U.S. government is expected to owe $10 trillion to bondholders. That amount is equal to 70% of the expected 2010 Gross Domestic Product (GDP), which measures our country's total output for a year. The debt may become more burdensome in the next decade unless something is done. The federal debt may increase for the following reasons:

- The financial crisis in 2008 led the government to "bail out" banks and other corporations that appeared to be failing. The bailout programs resulted in $1.5 trillion of extra spending in 2009 and 2010. The government has taken equity (common stock) stakes in companies that have been bailed out and will eventually try to sell some of the companies' common stock to recover some of the bailout money.
- The U.S. government will continue to spend heavily on the Iraq and Afghanistan wars. Total outlays for defense will be approximately $1 trillion in 2010, including related spending for the CIA, State Department, and other agency budgets. Defense outlays would decline significantly if the two wars end.
- Social service expenditures will escalate greatly in the next decade as "baby boomers" start to receive Social Security payments and as government-funded medical expenses increase for aging boomers.
- High deficits might result in much higher inflation, which would drive up the interest rate paid on U.S. government bonds. Of course, this would increase interest paid to debt holders. However, note that some respected economists think the nation's high deficits may not cause high inflation in coming years. They believe that the general level of prices will remain flat regardless of deficits.

A forecast is needed for U.S. budget debt owed in the next decade. Major sources of U.S. government revenues and expenditures are discussed next.

Revenues

Taxes on individual income have averaged slightly more than 8% of wages and salaries in recent years. Generally, wages and salaries are approximately 45% of GDP, and taxes on corporate net income account for about 2.5% of GDP.

Citizens and businesses pay a variety of taxes that fund U.S. social programs. Primarily, these taxes pay for Social Security, Medicare, and Medicaid. The total of such tax receipts is approximately 6.3% of GDP.

Other miscellaneous taxes and fees are usually about 1.3% of GDP.

Expenditures

Social Security, Medicare, Medicaid, and other social safety net expenditures are "entitlements" to American taxpayers. Unless Congress changes the law, the government must spend the money. These mandatory expenditures are expected to increase greatly in the next decade as the large baby boomer generation retires and requires more medical attention.

Military spending, which is classified as "discretionary" in the U.S. budget, has increased greatly in recent years. The Iraq and Afghanistan wars have been very expensive. Overall, military expenditures are expected to increase 3% a year, but if the two wars can be ended, military spending could be significantly reduced.

Bailout spending to ease the financial crisis began in 2009 and will end in 2010. After that, the government may be able to recover some of the bailout funding as equity stakes are liquidated in bailed-out companies. Recovered funds would be a source of revenue for the government.

All other U.S. spending, including that for commerce, education, space exploration, federal courts, and so on, is classified as "discretionary." This spending, on average, accounts for 3.7% of GDP.

U.S. government bonds have different maturities; some bonds are paid back after a year, some after 10 years, and some after 30 years. For budgeting purposes, analysts assume that 8% of outstanding U.S. debt will come due ("mature") and be paid back each year. Repayment of U.S. government debt is an expenditure.

Although in recent years interest rates have been low by historical standards, the level of debt is very high by any standard. Thus, interest on outstanding debt is expected to be a significant U.S. government expenditure in the coming decade.

U.S. Debt Outstanding

If the U.S. government spends more than its revenues, the United States must borrow the difference, which of course increases the amount of debt outstanding. On the other hand, if revenues exceed expenditures, the surplus can be used to reduce U.S. debt. In other words, debt could be repaid before its maturity date. In your analysis, you can assume that surpluses would be used to repay debt early.

At the end of 2010, the U.S. government's debt was expected to be $10 trillion, which would be 70% of 2010's GDP ($14.286 trillion). This percentage would be the highest since the World War II era. Budget analysts are worried about the high level of this ratio.

ASSIGNMENT 1: CREATING A SPREADSHEET FOR DECISION SUPPORT

In this assignment, you will produce a spreadsheet that models the U.S. deficit. Then, in Assignment 2, you will write a memorandum about your analysis and provide recommendations for dealing with the deficit. In Assignment 3, you will prepare and give an oral presentation of your analysis and recommendations.

First, you will create the spreadsheet model of the government's deficit position. The model covers the years from 2010 to 2020. This section helps you set up each of the following spreadsheet components before entering cell formulas:

- Constants
- Inputs
- Summary of key results
- Calculations
- Calculation of U.S. debt owed

A discussion of each section follows. *The spreadsheet skeleton is available for you to use; you can choose to type in it or not.* To access the spreadsheet skeleton, go to your data files, select Case 7, and then select **Deficit.xlsx**.

Constants Section

Your spreadsheet should include the constants shown in Figure 7-1. An explanation of each item follows the figure.

	A	B	C	D	E	F	G	H	I	J	K	L
1	U.S. BUDGET CALCULATOR, 2010-2020											
2												
3	CONSTANTS	2010	2011	2012	2013	2014	2015	2016	2017	2018	2019	2020
4	EXPECTED GROSS DOMESTIC PRODUCT	14286	14350	14400	15347	16293	17280	18211	19077	19909	20749	21617
5	EXPECTED INTEREST RATE ON FEDERAL DEBT	NA	0.040	0.045	0.045	0.050	0.055	0.055	0.050	0.050	0.050	0.050
6	EXPECTED MANDATORY SOCIAL SPENDING	NA	1600	1350	1900	1900	2025	2150	2300	2500	2600	2700

FIGURE 7-1 Constants section

- Expected Gross Domestic Product—Yearly GDP estimates have been provided by a respected government economist. These estimates are shown in multiples of billions.
- Expected Interest Rate on Federal Debt—The economist's interest rate estimates are shown for each year. These estimates assume that inflation will stay under control.
- Expected Mandatory Social Spending—The economist's estimates of expenditures are shown for Social Security, Medicare, and so on. The amounts increase greatly over time. These estimates are shown in multiples of billions.

Inputs Section

Your spreadsheet should include the inputs shown in Figure 7-2. An explanation of each item follows the figure. (An "NA" means that the cell address should not appear in a spreadsheet formula.)

	A	B	C	D	E	F	G	H	I	J	K	L
8	INPUTS	2010	2011	2012	2013	2014	2015	2016	2017	2018	2019	2020
9	PERCENTAGE BAILOUT RECOVERY IN YEAR	NA				NA	NA	NA	NA	NA	NA	NA
10	ZERO WARS 2015-2020? (YES/NO)		NA	NA	NA	NA	NA	NA	NA	NA	NA	NA
11	INCREASE IN SOCIAL OUTLAY TAXATION (.XX)		NA	NA	NA	NA	NA	NA	NA	NA	NA	NA
12	SERIOUS INFLATION? (YES/NO)		NA	NA	NA	NA	NA	NA	NA	NA	NA	NA

FIGURE 7-2 Inputs section

- Percentage Bailout Recovery in Year—Bailout expenditures will occur in 2009 and 2010. It is assumed that these bailouts will be recovered in 2011 through 2013, with no recoveries after 2013. For example, if you think that 25% of total bailout outlays will be recovered in 2011, 2012, and 2013, you would enter .25, .25, and .25 in the input cells. If you think that 30% will be recovered in 2011, but only 10% will be recovered in the next two years, your entries would be .30, .10, and .10.

- Zero Wars 2015-2020? (YES/NO)—The hope is that the Iraq and Afghanistan wars will end by 2014, and that no foreign wars will occur from 2015 to 2020. If you think this will happen, enter YES. If not, enter NO. (Do not use Y and N.)
- Increase in Social Outlay Taxation (.XX)—A variety of taxes are used to fund U.S. social programs. The primary taxes are for Social Security, Medicare, and Medicaid, which typically account for 6.3% of GDP. Because much higher outlays are expected to be needed to fund the social safety net, a higher tax rate may be needed. If you wanted to know the effects of increasing this overall rate by 5 percentage points, you would enter .05. With that entry, the model would use 11.3% as the tax rate for social service tax revenues (.063 + .050). If you expected no increase in the rate of taxation, you would enter zero.
- Serious Inflation? (YES/NO)—If you think inflation will escalate in coming years, enter YES. Otherwise, enter NO. This entry will affect the interest rate used to compute interest expense. (Do not use Y and N.)

Summary of Key Results Section

Your spreadsheet should include the results shown in Figure 7-3. An explanation of each item follows the figure.

	A	B	C	D	E	F	G	H	I	J	K	L
14	SUMMARY OF KEY RESULTS	2010	2011	2012	2013	2014	2015	2016	2017	2018	2019	2020
15	DEBT OWED -- END OF YEAR	NA										
16	PERCENTAGE OF GROSS DOMESTIC PRODUCT	NA										

FIGURE 7-3 Summary of Key Results section

- Debt Owed – End of Year—The amount of U.S. government debt owed at the end of each year is computed elsewhere in the spreadsheet and echoed here.
- Percentage of Gross Domestic Product—The percentage of debt owed to GDP at the end of each year is computed elsewhere in the spreadsheet and echoed here.

Calculations Section

You should calculate intermediate results that will be used in the projection that follows. Calculations, as shown in Figure 7-4, are based on inputs and (in some cases) on prior year-end values. You may need to use absolute addressing in some formulas. Values must be computed by cell formulas, and no cell formulas should reference a cell that contains an "NA" value. Save your work frequently. An explanation of each item follows the figure.

	A	B	C	D	E	F	G	H	I	J	K	L
18	CALCULATIONS	2010	2011	2012	2013	2014	2015	2016	2017	2018	2019	2020
19	MILITARY OUTLAY	1000										
20	SOCIAL OUTLAY TAX RATE	0.063	0.063									
21	SALARY AND WAGES	NA										
22	INDIVIDUAL INCOME TAX REVENUE	NA										
23	INTEREST RATE ON FEDERAL DEBT	NA	0.040									

FIGURE 7-4 Calculations section

- Military Outlay—This outlay was $1 trillion in 2010 (shown as 1000 in cell B19). This calculation consists of three parts. (a) Military outlays are expected to increase 3% each year from 2011 through 2014. In other words, each year's outlay would be 3% greater than the prior year's outlay. (b) The outlay in 2015 will fall to $600 billion if you indicate that no wars will occur in 2015 through 2020. However, if you indicate that wars will continue, 2015 military outlays will be 3% greater than 2014 military outlays. (c) Outlays from 2016 to 2020 are expected to increase 2% each year. In other words, each year's outlay would be 2% greater than the prior year's outlay (regardless of expectations for the two wars).

- Social Outlay Tax Rate—In 2011, the overall tax rate is expected to stay at 6.3%, as shown. After that, the rate for each year equals .063 plus your entry for the Increase in Social Outlay Taxation (.XX) variable in the Inputs section. For example, if you enter .02 for the variable, the social outlay tax rate in 2020 would be .083 (.063 + .020).
- Salary and Wages—The salary and wages earned by individual taxpayers in the year. The amount equals 45% (.45) of expected GDP in each year. Expected GDP values are taken from the Constants section.

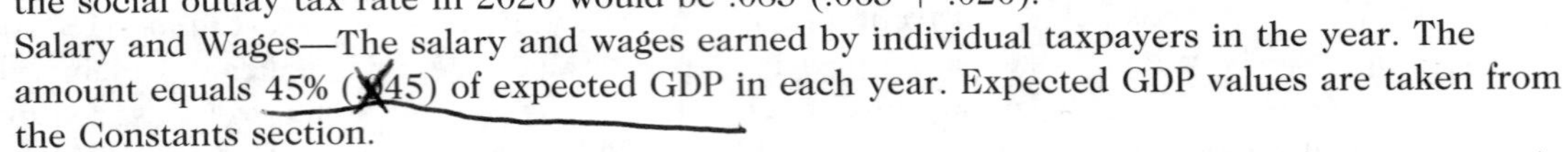

- Individual Income Tax Revenue—This revenue equals 8.2% (.082) of salary and wages earned in the year. The Salary and Wages value is taken from the Calculations section.

- Interest Rate on Federal Debt—The rate is expected to be 4% (.04) in 2011, as shown. Thereafter, if you do not expect serious inflation, the interest rate on federal debt equals the interest rate value shown in the Constants section. If you do expect serious inflation, the interest rate on federal debt equals the value shown in the Constants section plus 3% (.03). Note that your inflation expectation is a value from the Inputs section.

Calculation of U.S. Debt Owed Section

The forecast shows the difference between expected revenues and outlays in each year, followed by the calculation of debt owed and the ratio of debt owed to GDP in each year. Values must be computed by cell formulas. The spreadsheet sections look like those in Figures 7-5 and 7-6. An explanation of each item follows the figure.

	A	B	C	D	E	F	G	H	I	J	K	L
25	CALCULATION OF U.S. DEBT OWED	2010	2011	2012	2013	2014	2015	2016	2017	2018	2019	2020
26	REVENUE											
27	INDIVIDUAL INCOME TAXES	NA										
28	CORPORATE INCOME TAXES	NA										
29	SOCIAL SAFETY NET TAXES	NA										
30	BAILOUT RECOVERY	NA				0	0	0	0	0	0	0
31	OTHER RECEIPTS	NA										
32	TOTAL REVENUE	NA										
33	OUTLAYS											
34	MANDATORY SOCIAL	NA										
35	DISCRETIONARY -- MILITARY	NA										
36	DISCRETIONARY -- OTHER	NA										
37	8% REPAYMENT OF EXISTING DEBT	NA										
38	INTEREST ON DEBT	NA										
39	TOTAL OUTLAYS	NA										
40	TOTAL REVENUE LESS TOTAL OUTLAY	NA										

FIGURE 7-5 Calculation of revenues and outlays

- Individual Income Taxes—This amount has been calculated above and can be echoed here.
- Corporate Income Taxes—This amount equals 2.5% (.025) of the year's expected GDP.
- Social Safety Net Taxes—This amount equals the social outlay tax rate from the Calculations section times the year's expected GDP.
- Bailout Recovery—In most years this amount is zero, as shown. In 2011 through 2013, the amount equals the total of bailout outlays ($1.5 trillion) times the year's percentage bailout recovery. Each of the years from 2011 to 2013 has a Percentage Bailout Recovery value from the Inputs section. (In your formula you would use the number 1500 to represent the $1.5 trillion bailout value.)
- Other Receipts—This amount equals 1.3% (.013) of the year's expected GDP.
- Total Revenue—This yearly amount equals the sum of individual income taxes, corporate taxes, social safety net taxes, bailout recovery, and other receipts.
- Mandatory Social—Mandatory social spending outlays are shown in the Constants section and can be echoed here.
- Discretionary – Military—Military outlays are computed in the Calculations section and can be echoed here.
- Discretionary – Other—These amounts equal 3.7% (.037) times the year's expected GDP.
- 8% Repayment of Existing Debt—Eight percent (.08) times the debt owed at the beginning of the year.
- Interest on Debt—The interest rate on debt (from the Calculations section) times the debt owed at the beginning of the year.
- Total Outlays—The sum of mandatory social and military outlays, other spending, debt repayment, and interest on debt owed.
- Total Revenue Less Total Outlay—This amount equals total revenue minus total outlays. A positive number represents a surplus for the year. A negative number represents a deficit for the year.

The calculation of debt owed by the U.S. government is shown in Figure 7-6. Each item is explained after the figure.

	A	B	C	D	E	F	G	H	I	J	K	L
42	DEBT OWED -- BEGINNING OF YEAR	NA										
43	TOTAL REVENUE LESS TOTAL OUTLAY	NA										
44	DEBT OWED -- END OF YEAR	10000										
45	PERCENTAGE OF GROSS DOMESTIC PRODUCT	70.0%										

FIGURE 7-6 Calculation of debt owed

- Debt Owed – Beginning of Year—This amount equals the debt owed at the end of the prior year.
- Total Revenue Less Total Outlay—This amount is calculated above and can be echoed here. This amount is the net change in debt owed in the year.
- Debt Owed – End of Year—There are two possibilities, assuming that $10 trillion is owed at the beginning of the year. The first possibility is that a surplus is used to reduce debt. If total revenue minus total outlays equals $1 trillion, then debt owed at the end of the year equals $9 trillion ($10 trillion minus the $1 trillion surplus). The second possibility is that a deficit results in more debt. If total revenue minus total outlays equals negative $1 trillion, then debt owed at the end of the year equals $11 trillion ($10 trillion plus the $1 trillion deficit).
- Percentage of Gross Domestic Product—Debt owed at the end of the year divided by the year's expected GDP. (The 2010 percentage is shown as an example: $10 trillion divided by $14.286 expected GDP.)

ASSIGNMENT 2: USING THE SPREADSHEET FOR DECISION SUPPORT

You complete the case by using the spreadsheet to gather data needed to assist policy makers and by documenting recommendations in a memorandum.

Budget analysts want to consider three scenarios: (1) an optimistic scenario, which might result in a low percentage of debt owed to GDP; (2) a pessimistic scenario, which might result in a high percentage of debt owed to GDP; and (3) a so-so scenario, which presumably would yield results between the optimistic and pessimistic scenarios.

Recall that high amounts of bailout money might be recovered, increasing the possibility of surpluses. If low amounts of bailout money are recovered, the possibility of deficits would increase. Recall that ending the wars would reduce military expenditures, increasing the possibility of surpluses; continuing the wars would increase the possibility of deficits. Social tax rate increases would increase revenues, thus increasing the possibility of surpluses; low (or no) tax rate increases would increase the possibility of deficits. Serious inflation would be expected to increase interest rates and interest expenditures, thus increasing the possibility of deficits. On the other hand, low inflation would increase the possibility of low interest rates and interest expenditures, thus increasing the possibility of surpluses.

Assignment 2A: Using the Spreadsheet to Gather Data

You have built the spreadsheet to model the government's financial situation. For each of the three scenarios, you want to know the debt owed at the end of the year and the ratio of debt owed at the end of the year to the year's GDP.

Bailout recovery percentages are shown in input cells C9, D9, and E9. The expectation of two wars continuing is shown in cell B10, the social tax rate increase is shown in cell B11, and the expectation of serious inflation is shown in cell B12. The results cells in each case are the level of debt and percentage of debt owed at the end of 2015 and the end of 2019. The table in Figure 7-7 summarizes the input values for the three scenarios.

645K45

Scenario Name	C9	D9	E9	B10	B11	B12
Optimistic	.3	.3	.3	YES	.15	NO
Pessimistic	.1	.1	.1	NO	.01	YES
So-So	.2	.2	.2	NO	.05	NO

FIGURE 7-7 Summary of scenario input values

You will run "what-if" scenarios with the three sets of inputs using the Scenario Manager. See Tutorial C for tips on using the Scenario Manager.

Set up the three scenarios. This example shows how the Scenario Manager can handle noncontiguous input cell ranges of C10, D11, C12:F12 (cell addresses are arbitrary). The output cells are Debt Owed – End of Year and Percentage of Gross Domestic Product from the Summary of Key Results section for the years 2015 and 2020. When you finish, print the spreadsheet with the input for any of the scenarios, print the Scenario Manager summary sheet, and then save the spreadsheet file for the last time.

U.S. government budget officials are afraid of seeing the level of debt surpass 100% of GDP. They want to know what (if any) scenarios result in percentages below 100% in 2015 and 2020. For such situations, they want to know the policy implications. In other words, should the wars be ended and should taxes be raised to achieve percentages below 100%?

Assignment 2B: Documenting Your Recommendations in a Memorandum

Use Microsoft Word to write a brief memorandum to the director of the Congressional Budget Office. The memo should summarize your findings and your recommendations. Observe the following requirements:

- Set up your memorandum as described in Tutorial E.
- In the first paragraph, briefly summarize the key revenues and expenditures.
- State your findings. Which scenarios result in debt percentages over 100%, and which scenarios result in debt percentages below 100%?
- If results differ from 2015 to 2020, describe the differences. If results are consistent, state that fact.
- State the policy implications for scenarios that result in debt percentages below 100%. In the model, did the two wars need to end, and by what percentage did taxes need to be raised to achieve a percentage less than 100?
- Support your statements graphically, as your instructor requires. If you used the Scenario Manager, your instructor may ask you to return to Excel and copy the Scenario Manager summary sheet results into the memorandum. (See Tutorial C for details on this procedure.) Your instructor might also ask you to make a summary table in Word based on the Scenario Manager summary sheet results. (This procedure is described in Tutorial E.) Your table should resemble the format shown in Figure 7-8.

	Optimistic	Pessimistic	So-So
Bailout Recovery %, 2011			
Bailout Recovery %, 2012			
Bailout Recovery %, 2013			
No wars, 2015-2020? (Y/N)			
Social taxes increase (.XX)			
Serious inflation? (Y/N)			
Debt owed at end of 2015			
Debt owed to GDP%—2015			
Debt owed at end of 2020			

FIGURE 7-8 Format of table to insert in memorandum

ASSIGNMENT 3: GIVING AN ORAL PRESENTATION

Your instructor may ask you to present your analysis and recommendations in an oral presentation. Prepare to explain your analysis and recommendations to a group of government budget analysts in 10 minutes or less. Use visual aids or handouts that you think are appropriate. See Tutorial F for tips on preparing and giving an oral presentation.

DELIVERABLES

Assemble the following deliverables for your instructor:

1. Printout of your memorandum
2. Spreadsheet printouts
3. Electronic media such as a USB key or CD that contains your Word memo and Excel spreadsheet file

Staple the printouts together with the memorandum on top. If you have more than one .xlsx file on your electronic media, write your instructor a note that identifies your model's .xlsx file.

PART 3

DECISION SUPPORT CASES USING THE EXCEL SOLVER

BUILDING A DECISION SUPPORT SYSTEM USING THE EXCEL SOLVER

Decision support systems (DSS) help people make decisions. (The nature of DSS programs is discussed in Tutorial C.) Tutorial D teaches you how to use the Solver, one of the Excel built-in decision support tools.

For some business problems, decision makers want to know the best, or optimal, solution. Usually that means maximizing a variable (for example, net income) or minimizing another variable (for example, total costs). This optimization is subject to constraints, which are rules that must be observed when solving a problem. The Solver computes answers to such optimization problems.

This tutorial has four sections:

1. **Using the Excel Solver**: In this section, you'll learn how to use the Solver in decision making. As an example, you use the Solver to create a production schedule for a sporting goods company. This schedule is called the base case.
2. **Extending the Example**: In this section, you'll test what you've learned about using the Solver as you modify the sporting goods company's production schedule. This is called the extension case.
3. **Using the Solver on a New Problem**: In this section, you'll use the Solver on a new problem.
4. **Troubleshooting the Solver**: In this section, you'll learn how to overcome problems you might encounter when using the Solver.

NOTE

Tutorial C offers some guidance on basic Excel concepts, such as formatting cells and using functions such as =IF(). Refer to Tutorial C for a review of those topics.

USING THE EXCEL SOLVER

Suppose a company must set a production schedule for its various products, each of which has a different profit margin (selling price, less costs). At first, you might assume the company will maximize production of all profitable products to maximize net income. However, a company typically cannot make and sell an unlimited number of its products because of constraints.

One constraint affecting production is the shared resource problem. For example, several products in a manufacturer's line might require the same raw materials, which are in limited supply. Similarly, the manufacturer might require the same machines to make several of its products. In addition, there might be a limited pool of skilled workers available to make the products.

In addition to production constraints, sometimes management's policies impose constraints. For example, management might decide that the company must have a broader product line. As a consequence, a certain production quota for several products must be met, regardless of profit margins.

Thus, management must find a production schedule that will maximize profit given the constraints. Optimization programs like the Solver look at each combination of products, one after the other, ranking each combination by profitability. Then the program reports the most profitable combination.

To use the Solver, you'll set up a model of the problem, including the factors that can vary, the constraints on how much they can vary, and the goal—that is, the value you are trying to maximize (usually net income) or minimize (usually total costs). The Solver then computes the best solution.

Setting Up a Spreadsheet Skeleton

Suppose your company makes two sporting goods products—basketballs and footballs. Assume you will sell all of the balls you produce. To maximize net income, you want to know how many of each kind of ball to make in the coming year.

Making each kind of ball requires a certain (and different) number of hours, and each ball has a different raw materials cost. Because you have only a limited number of workers and machines, you can devote a maximum of 40,000 hours to production, which is a shared resource. You do not want that resource to be idle, however. Downtime should be no more than 1,000 hours in a year, so machines should be used for at least 39,000 hours.

Marketing executives say that you cannot make more than 60,000 basketballs and cannot make fewer than 30,000. Furthermore, they say that you must make at least 20,000 footballs but not more than 40,000. Marketing also says that the ratio of basketballs to footballs produced should be between 1.5 and 1.7—that is, more basketballs than footballs, but within limits.

What would be the best production plan? This problem has been set up in the Solver. The spreadsheet sections are discussed in the pages that follow.

AT THE KEYBOARD

Start by saving the blank spreadsheet as **SPORTS1.xlsx**. Then enter the skeleton and formulas as they are discussed.

CHANGING CELLS Section

The CHANGING CELLS section contains the variables the Solver is allowed to change while it looks for the solution to the problem. Figure D-1 shows the skeleton of this spreadsheet section and the values you should enter. An analysis of the line items follows the figure.

	A	B	C	D	E
1	**SPORTING GOODS EXAMPLE**				
2	CHANGING CELLS				
3	NUMBER OF BASKETBALLS	1			
4	NUMBER OF FOOTBALLS	1			

FIGURE D-1 CHANGING CELLS section

- The changing cells are for the number of basketballs and footballs to be made and sold. The changing cells are like input cells, except that the Solver (not you) plays "what if" with the values, trying to maximize or minimize some value (in this case, maximize net income).
- Some number should be put in the changing cells each time, before the Solver is run. It is customary to put the number 1 in the changing cells (as shown). The Solver will change these values when the program is run.

CONSTANTS Section

Your spreadsheet also should have a section for values that will not change. Figure D-2 shows a skeleton of the CONSTANTS section and the values you should enter. A discussion of the line items follows the figure.

NOTE

You should format cells in the constants range to two decimal places.

	A	B	C	D	E
6	**CONSTANTS**				
7	BASKETBALL SELLING PRICE	14.00			
8	FOOTBALL SELLING PRICE	11.00			
9	TAX RATE	0.28			
10	NUMBER OF HOURS TO MAKE A BASKETBALL	0.50			
11	NUMBER OF HOURS TO MAKE A FOOTBALL	0.30			
12	COST OF LABOR -- 1 MACHINE HOUR	10.00			
13	COST OF MATERIALS -- 1 BASKETBALL	2.00			
14	COST OF MATERIALS -- 1 FOOTBALL	1.25			

FIGURE D-2 CONSTANTS section

- The SELLING PRICE for a single basketball and a single football is shown.
- The TAX RATE is the rate applied to income before taxes to compute income tax expense.
- The NUMBER OF HOURS needed to make a basketball and a football is shown. Note that a ball-making machine can produce two basketballs in an hour.
- COST OF LABOR: A ball is made by a worker using a ball-making machine. A worker is paid $10 for each hour he or she works at a machine.
- COST OF MATERIALS: The costs of raw materials for a basketball and football are shown.

Notice that the profit margins (selling price, less costs of labor and materials) for the two products are not the same. They have different selling prices and different inputs (raw materials and hours to make), and the inputs have different costs per unit. Also note that you cannot tell from the data how many hours of the shared resource (machine hours) will be devoted to basketballs and how many will be devoted to footballs, because you don't know in advance how many basketballs and footballs will be made.

CALCULATIONS Section

In the CALCULATIONS section, you will calculate intermediate results that (1) will be used in the spreadsheet body and/or (2) will be used as constraints. Before entering formulas, format the Calculations range cells for two decimal places (cell-formatting directions were given in Tutorial C). Figure D-3 shows the skeleton and formulas you should enter. A discussion of the cell formulas follows the figure.

NOTE

Cell widths are changed here merely to show the formulas—you need not change the width.

	A	B	C	D
16	**CALCULATIONS**			
17	RATIO OF BASKETBALLS TO FOOTBALLS	=B3/B4		
18	TOTAL BASKETBALL HOURS USED	=B3*B10		
19	TOTAL FOOTBALL HOURS USED	=B4*B11		
20	TOTAL MACHINE HOURS USED (BB + FB)	=B18+B19		

FIGURE D-3 CALCULATIONS section cell formulas

- RATIO OF BASKETBALLS TO FOOTBALLS: This number (cell B17) will be needed in a constraint.
- TOTAL BASKETBALL HOURS USED: The number of machine hours needed to make all basketballs (B3 * B10) is computed in cell B18. Cell B10 has the constant for the hours needed to make one basketball. Cell B3 (a changing cell) has the number of basketballs made. (Currently, this cell shows one ball, but that number will change when the Solver works on the problem.)
- TOTAL FOOTBALL HOURS USED: The number of machine hours needed to make all footballs is calculated similarly in cell B19.
- TOTAL MACHINE HOURS USED (BB + FB): The number of hours needed to make both kinds of balls (cell B20) will be used in constraints; this value is the sum of the hours just calculated for footballs and basketballs.

Notice that constants in the Excel cell formulas in Figure D-3 are referred to by their cell addresses. Use the cell address of a constant rather than hard-coding a number in the Excel expression. If the number must be changed later, you will have to change it only in the CONSTANTS section cell, not in every cell formula in which you used the value.

Notice that you do not calculate the amounts in the changing cells (here, the number of basketballs and footballs to produce). The Solver will compute those numbers. Also, notice that you can use the changing cell addresses in your formulas. When you do that, you assume the Solver has put the optimal values in each changing cell; your expression makes use of those numbers.

Figure D-4 shows the calculated values after Excel evaluates the cell formulas (with 1s in the changing cells).

	A	B	C	D	E
16	CALCULATIONS				
17	RATIO OF BASKETBALLS TO FOOTBALLS	1.00			
18	TOTAL BASKETBALL HOURS USED	0.50			
19	TOTAL FOOTBALL HOURS USED	0.30			
20	TOTAL MACHINE HOURS USED (BB + FB)	0.80			

FIGURE D-4 CALCULATIONS section cell values

INCOME STATEMENT Section

The target value is calculated in the spreadsheet body in the INCOME STATEMENT section. This is the value that the Solver is expected to maximize or minimize. The spreadsheet body can take any form. In this textbook's Solver cases, the spreadsheet body will be an income statement. Figure D-5 shows the skeleton and formulas that you should enter. A discussion of the line-item cell formulas follows the figure.

NOTE

Income statement cells were formatted for two decimal places.

	A	B	C
22	INCOME STATEMENT		
23	BASKETBALL REVENUE (SALES)	=B3*B7	
24	FOOTBALL REVENUE (SALES)	=B4*B8	
25	TOTAL REVENUE	=B23+B24	
26	BASKETBALL MATERIALS COST	=B3*B13	
27	FOOTBALL MATERIALS COST	=B4*B14	
28	COST OF MACHINE LABOR	=B20*B12	
29	TOTAL COST OF GOODS SOLD	=SUM(B26:B28)	
30	INCOME BEFORE TAXES	=B25-B29	
31	INCOME TAX EXPENSE	=IF(B30<=0,0,B30*B9)	
32	NET INCOME AFTER TAXES	=B30-B31	

FIGURE D-5 INCOME STATEMENT section cell formulas

- REVENUE (cells B23 and B24) equals the number of balls times the respective unit selling price. The number of balls is in the changing cells, and the selling prices are constants.
- TOTAL REVENUE is the sum of basketball and football revenue.
- MATERIALS COST (cells B26 and B27) follows a similar logic: number of units times unit cost.
- COST OF MACHINE LABOR is the calculated number of machine hours times the hourly labor rate for machine workers.
- TOTAL COST OF GOODS SOLD is the sum of the cost of materials and the cost of labor.

This is the logic of income tax expense: If INCOME BEFORE TAXES is less than or equal to zero, the tax is zero; otherwise, the income tax expense equals the tax rate times income before taxes. An =IF() statement is needed in cell B31.

Excel evaluates the formulas. Figure D-6 shows the results (assuming 1s in the changing cells).

	A	B	C	D	E
22	**INCOME STATEMENT**				
23	BASKETBALL REVENUE (SALES)	14.00			
24	FOOTBALL REVENUE (SALES)	11.00			
25	TOTAL REVENUE	25.00			
26	BASKETBALL MATERIALS COST	2.00			
27	FOOTBALL MATERIALS COST	1.25			
28	COST OF MACHINE LABOR	8.00			
29	TOTAL COST OF GOODS SOLD	11.25			
30	INCOME BEFORE TAXES	13.75			
31	INCOME TAX EXPENSE	3.85			
32	NET INCOME AFTER TAXES	9.90			

FIGURE D-6 INCOME STATEMENT section cell values

Constraints

Constraints are rules that the Solver must observe when computing the optimal answer to a problem. Constraints need to refer to calculated values, or to values in the spreadsheet body. Therefore, you must build those calculations into the spreadsheet design so they are available to your constraint expressions. (There is no section on the face of the spreadsheet for constraints. You'll use a separate window to enter constraints.)

Figure D-7 shows the English and Excel expressions for the basketball and football production problem constraints. A discussion of the constraints follows the figure.

Expression in English	Excel Expression
TOTAL MACHINE HOURS >= 39000	B20 >= 39000
TOTAL MACHINE HOURS <= 40000	B20 <= 40000
MIN BASKETBALLS = 30000	B3 >= 30000
MAX BASKETBALLS = 60000	B3 <= 60000
MIN FOOTBALLS = 20000	B4 >= 20000
MAX FOOTBALLS = 40000	B4 <= 40000
RATIO BBs TO FBs-MIN = 1.5	B17 >= 1.5
RATIO BBs TO FBs-MAX = 1.7	B17 <= 1.7
NET INCOME MUST BE POSITIVE	B32 >= 0

FIGURE D-7 Solver constraint expressions

- As shown in Figure D-7, notice that a cell address in a constraint expression can be a cell address in the CHANGING CELLS section, a cell address in the CONSTANTS section, a cell address in the CALCULATIONS section, or a cell address in the spreadsheet body.
- You'll often need to set minimum and maximum boundaries for variables. For example, the number of basketballs (MIN and MAX) varies between 30,000 and 60,000.
- Often a boundary value is zero because you want the Solver to find a non-negative result. For example, here you want only answers that yield a positive net income. You tell the Solver that the amount in the net income cell must equal or exceed zero so the Solver does not find an answer that produces a loss.
- Machine hours must be shared between the two kinds of balls. The constraints for the shared resource are B20 >= 39000 and B20 <= 40000, where cell B20 shows the total hours used to make both basketballs and footballs. The shared resource constraint seems to be the most difficult kind of constraint for students to master when learning the Solver.
- Marketing wants some product balance, so ratios of basketballs to footballs must be in the constraints.

Running the Solver: Mechanics

To set up the Solver, you must tell the Solver:

- The cell address of the "target" variable that you are trying to maximize (or minimize, as the case may be)
- The changing cell addresses
- The expressions for the constraints

The Solver will record its answers in the changing cells and on a separate sheet.

Beginning to Set Up the Solver

AT THE KEYBOARD

To start setting up the Solver, first select the Data tab. In the Analysis group, select the Solver. The first thing you see is a Solver Parameters window, as shown in Figure D-8. Use the Solver Parameters window to specify the target cell, the changing cells, and the constraints. (If you do not see the Solver tool in the Analysis group, you should try to install it. Click the Office button, then click Excel Options, click Add-Ins, click the Go button for Manage: Excel Add-ins, select the check box for Solver Add-in, and then click OK. When prompted to install, click Yes.)

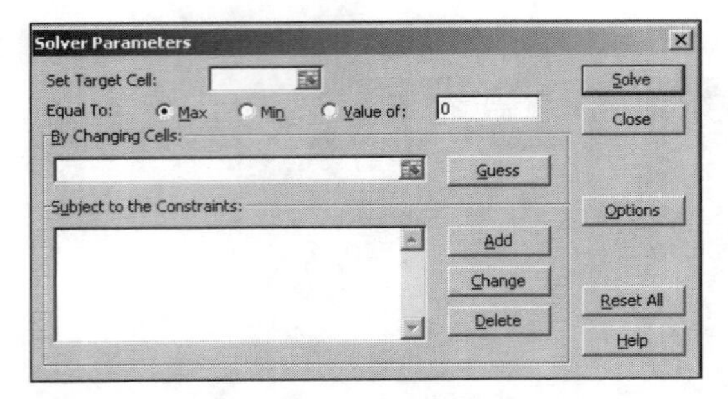

FIGURE D-8 Solver Parameters window

Setting the Target Cell

To set a target cell, use the following procedure:

1. Click the Set Target Cell box and enter B32 (net income).
2. Accept Max, the default.
3. Enter a zero for no desired net income value (Value of). Do *not* press Enter when you finish. You'll navigate within this window by clicking in the next input box.

Figure D-9 shows the entry of data in the Set Target Cell box.

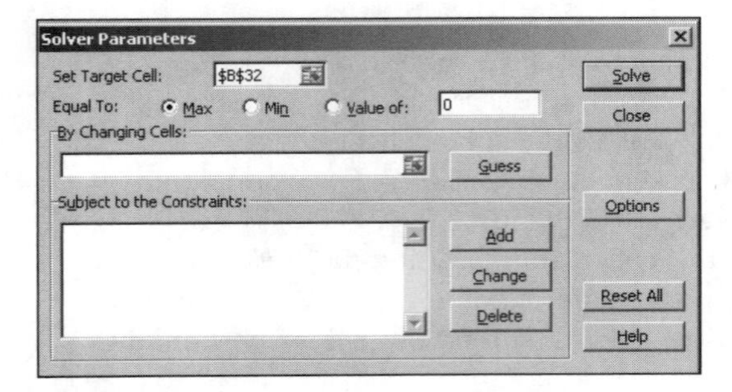

FIGURE D-9 Entering data in the Set Target Cell box

When you enter the cell address, the Solver may put in dollar signs, as if for absolute addressing. Ignore them—do not try to delete them.

Setting the Changing Cells

The changing cells are the cells for the balls, which are in the range of cells B3:B4. Click the By Changing Cells box and enter B3:B4, as shown in Figure D-10. (Do *not* press Enter.)

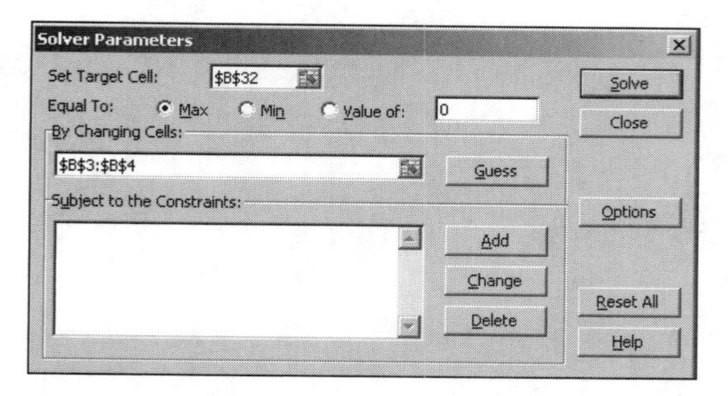

FIGURE D-10 Entering data in the By Changing Cells box

Entering Constraints

You are now ready to enter the constraint formulas one by one. To start, click the Add button. As shown in Figure D-11, you'll see the Add Constraint window (shown here with the minimum basketball production constraint entered).

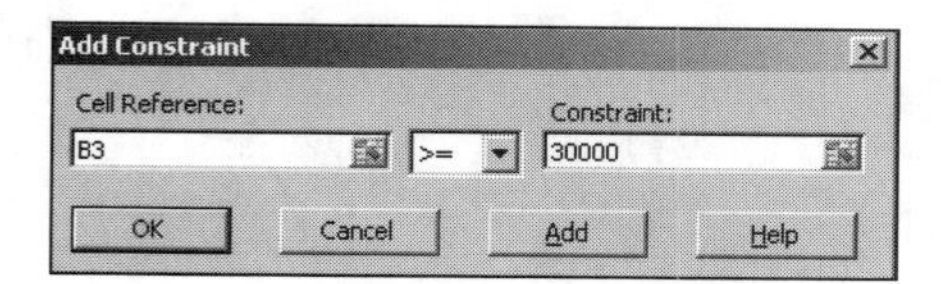

FIGURE D-11 Entering data in the Add Constraint window

You should note the following about entering constraints:

- To enter a constraint expression, do four things: (1) Type the variable's cell address in the left Cell Reference input box, (2) select the operator (<=, =, or >=) in the smaller middle box, (3) enter the expression's right-side value, which is either a raw number or the cell address of a value, into the Constraint box, and (4) click Add to enter the constraint into the program. If you change your mind about the expression and do not want to enter it, click Cancel.
- The minimum basketballs constraint is B3 >= 30000. Enter that constraint now. (Later, the Solver may put an equal sign in front of the 30000 and dollar signs in the cell reference.)
- After entering the constraint formula, click the Add button. Doing so puts the constraint into the Solver model. It also leaves you in the Add Constraint window, allowing you to enter other constraints. You should enter those now. See Figure D-7 for the logic.
- When you're done entering constraints, click the Cancel button. That takes you back to the Solver Parameters window.

Referring again to Figure D-11, you should not put an expression in the Cell Reference box. For example, the constraint for the minimum basketball-to-football ratio is B3/B4 >= 1.5. You should not put =B3/B4 in the Cell Reference box. The ratio is computed in the CALCULATIONS section of the spreadsheet (in cell B17). When adding that constraint, enter B17 in the Cell Reference box. (Although you are allowed to put an expression in the Constraint box, this technique is not recommended and is not shown here.)

After entering all of the constraints, you'll be back at the Solver Parameters window. You will see that the constraints have been entered into the program. Not all constraints will show, due to the size of the box. The top part of the box's constraints area looks like the portion of the spreadsheet shown in Figure D-12.

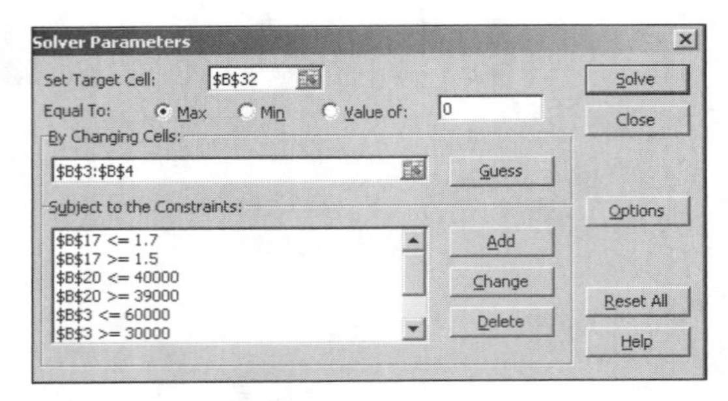

FIGURE D-12 A portion of the constraints entered in the Solver Parameters window

Using the scroll arrow, reveal the rest of the constraints, as shown in Figure D-13.

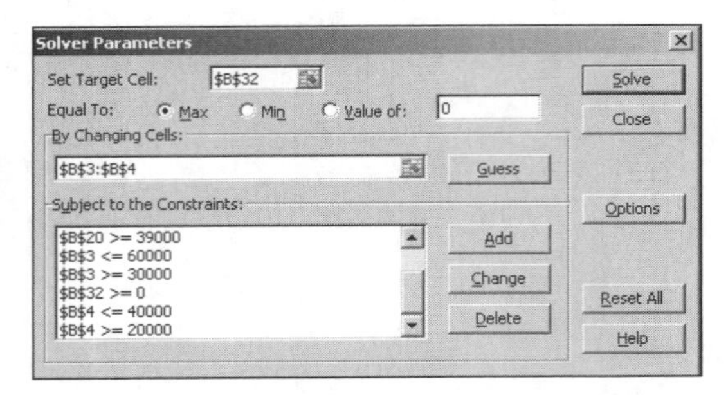

FIGURE D-13 Remainder of constraints entered in the Solver Parameters window

Computing the Solver's Answer

To have the Solver calculate answers, click Solve in the upper-right corner of the Solver Parameters window. The Solver does its work in the background—you do not see the internal calculations. Then the Solver gives you a Solver Results window, as shown in Figure D-14.

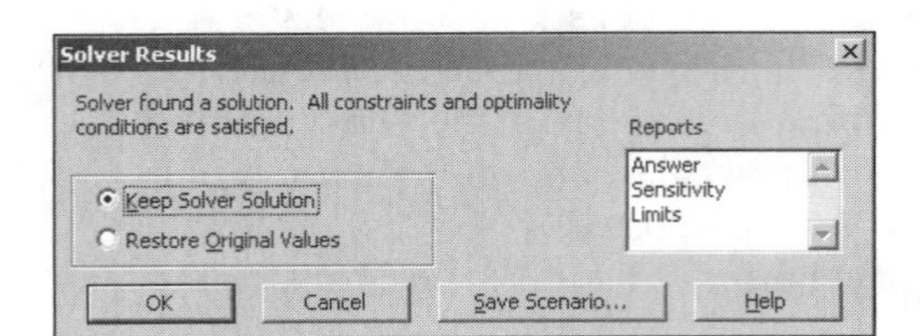

FIGURE D-14 Solver Results window

In the Solver Results window, the Solver tells you that it has found a solution and that the optimality conditions were met. That is a very important message—you should always check for it. It means that an answer was found and the constraints were satisfied.

By contrast, your constraints might be such that the Solver cannot find an answer. For example, suppose you had a constraint that said, in effect, "Net income must be at least a billion dollars." That amount cannot be reached, given so few basketballs and footballs and the prices. The Solver would report that no answer is feasible. The Solver may find an answer by ignoring some constraints. The Solver would tell you that, too. In either case, there would be something wrong with your model, and you would need to rework it.

There are two ways to see your answers. One way is to click OK, which lets you see the new changing cell values. A more formal (and complete) way is to click Answer in the Reports box and then click OK. That puts detailed results into a new sheet in your Excel book. The new sheet is called an Answer Report. All Answer Reports are numbered sequentially as you run the Solver.

To see the Answer Report, click its tab, as shown in Figure D-15 (here, Answer Report 1).

26	$B$4	NUMBER OF FOOTBALLS	38095.2442	$B$4>=20000	Not Binding
27	$B$4	NUMBER OF FOOTBALLS	38095.2442	$B$4<=40000	Not Binding
28					
29					
30					
31					
32					
33					
34					
35					
36					
37					

Answer Report 1 | Sheet1 | Sheet2 | Sheet3

FIGURE D-15 Answer Report Sheet tab

The top portion of the report is shown in Figure D-16.

	A	B	C	D	E	F
1	**Microsoft Excel 12.0 Answer Report**					
2	**Worksheet: [SPORTS1.xlsx]Sheet1**					
3						
4						
5						
6	Target Cell (Max)					
7		**Cell**	**Name**	**Original Value**	**Final Value**	
8		$B$32	NET INCOME AFTER TAXES	9.90	473142.87	
9						
10						
11	Adjustable Cells					
12		**Cell**	**Name**	**Original Value**	**Final Value**	
13		$B$3	NUMBER OF BASKETBALLS	1	57142.85348	
14		$B$4	NUMBER OF FOOTBALLS	1	38095.2442	

FIGURE D-16 Top portion of the Answer Report

Figure D-17 shows the remainder of the Answer Report.

	A	B	C	D	E	F
17	Constraints					
18		**Cell**	**Name**	**Cell Value**	**Formula**	**Status**
19		$B$17	RATIO OF BASKETBALLS TO FOOTBALLS	1.50	$B$17<=1.7	Not Binding
20		$B$17	RATIO OF BASKETBALLS TO FOOTBALLS	1.50	$B$17>=1.5	Binding
21		$B$20	TOTAL MACHINE HOURS USED (BB + FB)	40000.00	$B$20<=40000	Binding
22		$B$20	TOTAL MACHINE HOURS USED (BB + FB)	40000.00	$B$20>=39000	Not Binding
23		$B$32	NET INCOME AFTER TAXES	473142.87	$B$32>=0	Not Binding
24		$B$3	NUMBER OF BASKETBALLS	57142.85348	$B$3>=30000	Not Binding
25		$B$3	NUMBER OF BASKETBALLS	57142.85348	$B$3<=60000	Not Binding
26		$B$4	NUMBER OF FOOTBALLS	38095.2442	$B$4>=20000	Not Binding
27		$B$4	NUMBER OF FOOTBALLS	38095.2442	$B$4<=40000	Not Binding

FIGURE D-17 Remainder of the Answer Report

At the beginning of this tutorial, the changing cells had a value of 1, and the income was $9.90 (Original Value). The optimal solution values (Final Value) also are shown: $473,142.87 for net income (the target) and 57,142.85 basketballs and 38,095.24 footballs for the changing (adjustable) cells. (Of course, you cannot make a part of a ball. The Solver can be asked to find only integer solutions; that technique is discussed later in this tutorial.)

The report also shows detail for the constraints: the constraint expression and the value that the variable has in the optimal solution. *Binding* means the final answer caused the Solver to bump up against the constraint. For example, the maximum number of machine hours was 40,000, which is the value the Solver used to find the answer.

Not binding means the reverse. A better word for *binding* might be *constraining*. For example, the 60,000 maximum basketball limit did not constrain the Solver.

The procedures used to change (edit) or delete a constraint are discussed later in this tutorial.

Use the Office button to print the worksheets (Answer Report and Sheet1). Save the Excel file (Save). Then use Save As to make a new file called **SPORTS2.xlsx** to be used in the next section of this tutorial.

EXTENDING THE EXAMPLE

Next, you'll modify the sporting goods spreadsheet. Suppose management wants to know what net income would be if certain constraints were changed. In other words, management wants to play "what if" with certain base case constraints. The resulting second case is called the extension case. Here are some changes to the original base case conditions:

- Assume maximum production constraints will be removed.
- Similarly, the basketball-to-football production ratios (1.5 and 1.7) will be removed.
- There still will be minimum production constraints at some low level. Assume at least 30,000 basketballs and 30,000 footballs will be produced.
- The machine-hours shared resource imposes the same limits as previously.
- A more ambitious profit goal is desired. The ratio of net income after taxes to total revenue should be greater than or equal to .33. This constraint will replace the constraint calling for profits greater than zero.

AT THE KEYBOARD

Begin by putting 1s in the changing cells. You need to compute the ratio of net income after taxes to total revenue. Enter that formula in cell B21. (The formula should have the net income after taxes cell address in the numerator and the total revenue cell address in the denominator.) In the extension case, the value of this ratio for the Solver's optimal answer must be at least .33. Click the Add button and enter that constraint.

Then in the Solver Parameters window, constraints that are no longer needed are highlighted (click to select) and deleted (click the Delete button). Do that for the net income >= 0 constraint, the maximum football and basketball constraints, and the basketball-to-football ratio constraints.

The minimum football constraint must be modified, not deleted. Select that constraint, then click Change to open the Add Constraint window. Edit the constraint so 30,000 is the lower boundary.

When you finish with the constraints, your Solver Parameters window should look like the one shown in Figure D-18.

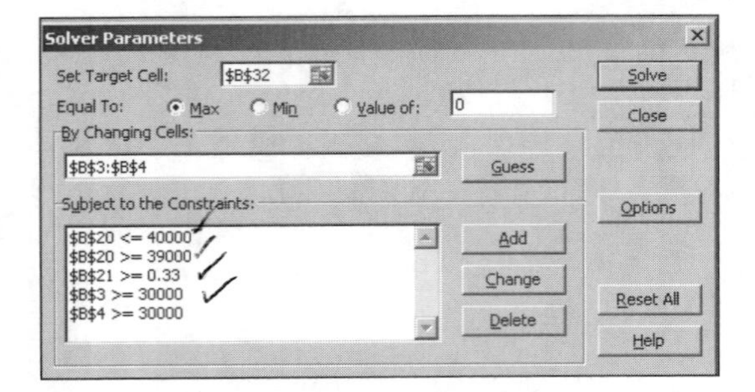

FIGURE D-18 Extension case Solver Parameters window

You can tell the Solver to solve for integer values. Here, cells B3 and B4 should be whole numbers. You use the Int constraint to do that. Figure D-19 shows how to enter the Int constraint.

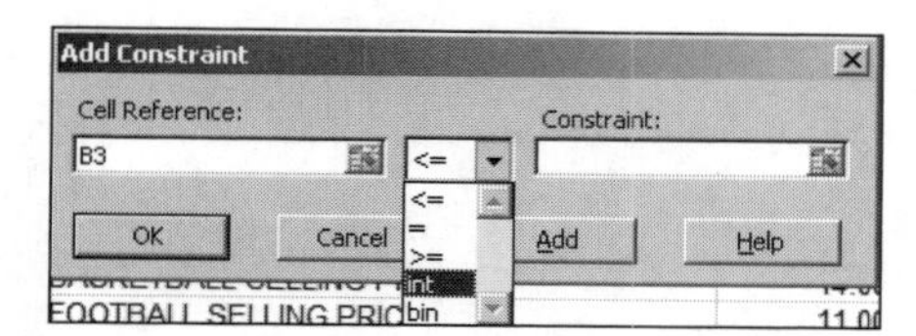

FIGURE D-19 Entering the Int constraint

Make those constraints for the changing cells. Your constraints should now look like the beginning portion of those shown in Figure D-20.

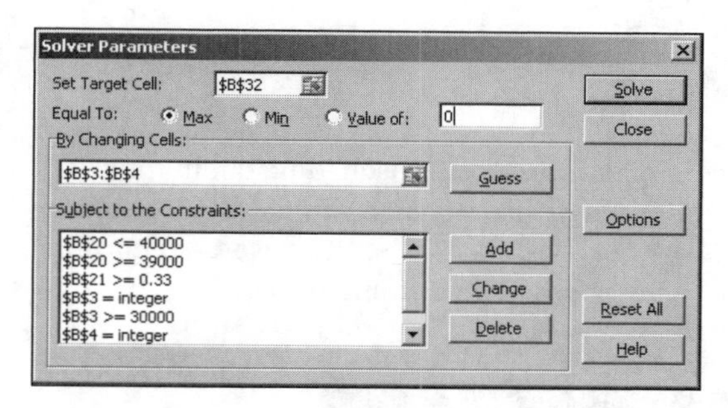

FIGURE D-20 Portion of extension case constraints

Scroll to see the remainder of the constraints, as shown in Figure D-21.

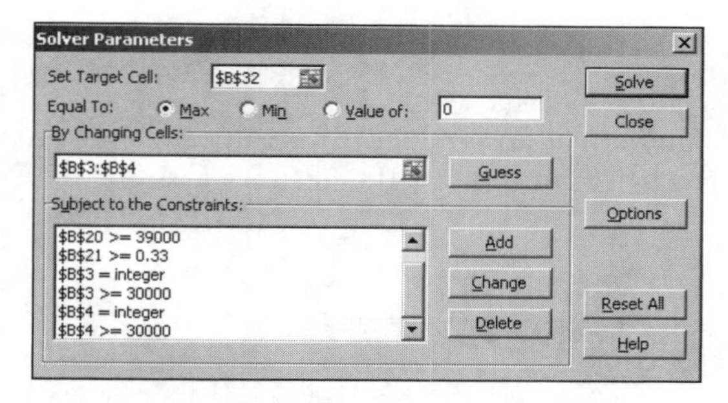

FIGURE D-21 Remainder of extension case constraints

The constraints are now only for the minimum production levels, the ratio of net income after taxes to total revenue, the machine-hours shared resource, and whole number output. When the Solver is run, the values in the Answer Report look like those shown in Figure D-22.

A	B	C	D	E	F
Target Cell (Max)					
	Cell	Name	Original Value	Final Value	
	$B$32	NET INCOME AFTER TAXES	9.90	556198.38	
Adjustable Cells					
	Cell	Name	Original Value	Final Value	
	$B$3	NUMBER OF BASKETBALLS	1	30000	
	$B$4	NUMBER OF FOOTBALLS	1	83333	
Constraints					
	Cell	Name	Cell Value	Formula	Status
	$B$20	TOTAL MACHINE HOURS USED (BB + FB)	39999.90	$B$20>=39000	Not Binding
	$B$21	RATIO OF NET INCOME TO TOTAL REVENUE	0.416109655	$B$21>=0.33	Not Binding
	$B$20	TOTAL MACHINE HOURS USED (BB + FB)	39999.90	$B$20<=40000	Not Binding
	$B$3	NUMBER OF BASKETBALLS	30000	$B$3>=30000	Binding
	$B$4	NUMBER OF FOOTBALLS	83333	$B$4>=30000	Not Binding
	$B$3	NUMBER OF BASKETBALLS	30000	$B$3=integer	Binding
	$B$4	NUMBER OF FOOTBALLS	83333	$B$4=integer	Binding

FIGURE D-22 Extension case Answer Report

The extension case answer differs from the base case answer. Which production schedule should management use: the one that has maximum production limits or the one that has no such limits? That question is posed to get you to think about the purpose of using a DSS program. Two scenarios, the base case and the extension case, were modeled in the Solver. The very different answers are shown in Figure D-23.

	Base Case	Extension Case
Basketballs	57,143	30,000
Footballs	38,095	83,333

FIGURE D-23 The Solver's answers for the two cases

Can you use this output alone to decide how many of each kind of ball to produce? No, you cannot. You also must refer to the "Target," which in this case is net income. Figure D-24 shows the answers with net income target data.

	Base Case	Extension Case
Basketballs	57,143	30,000
Footballs	38,095	83,333
Net Income	$473,143	$556,198

FIGURE D-24 The Solver's answers for the two cases—with target data

Viewed this way, the extension case production schedule looks better because it gives you a higher target net income.

At this point, you should save the SPORTS2.xlsx file and then close it (Office button—Close).

USING THE SOLVER ON A NEW PROBLEM

Here is a short problem that will let you find out what you have learned about the Excel Solver.

Setting Up the Spreadsheet

Assume you run a shirt-manufacturing company. You have two products: (1) polo-style T-shirts and (2) dress shirts with button-down collars. You must decide how many T-shirts and how many button-down shirts to make. Assume you'll sell every shirt you make.

AT THE KEYBOARD

Start a new file called **SHIRTS.xlsx**. Set up a Solver spreadsheet to handle this problem.

CHANGING CELLS Section

Your changing cells should look like those shown in Figure D-25.

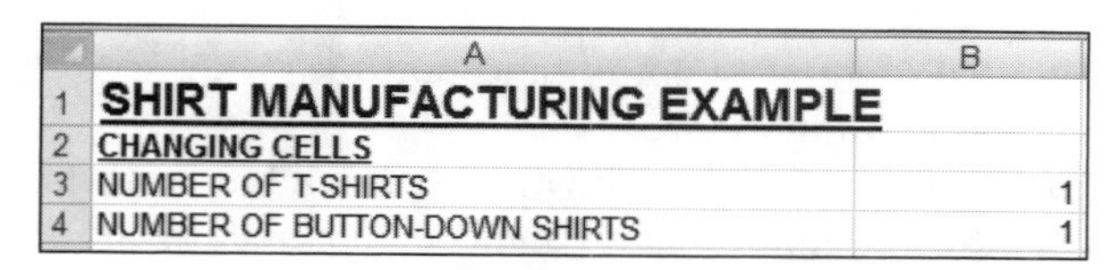

	A	B
1	SHIRT MANUFACTURING EXAMPLE	
2	CHANGING CELLS	
3	NUMBER OF T-SHIRTS	1
4	NUMBER OF BUTTON-DOWN SHIRTS	1

FIGURE D-25 Shirt manufacturing changing cells

CONSTANTS Section

Your spreadsheet should contain the constants shown in Figure D-26. A discussion of constant cells (and some of your company's operations) follows the figure.

	A	B
6	CONSTANTS	
7	TAX RATE	0.28
8	SELLING PRICE: T-SHIRT	8.00
9	SELLING PRICE: BUTTON-DOWN SHIRT	36.00
10	VARIABLE COST TO MAKE: T-SHIRT	2.50
11	VARIABLE COST TO MAKE: BUTTON-DOWN SHIRT	14.00
12	COTTON USAGE (LBS): T-SHIRT	1.50
13	COTTON USAGE (LBS): BUTTON-DOWN SHIRT	2.50
14	TOTAL COTTON AVAILABLE (LBS)	13000000
15	BUTTONS PER T-SHIRT	3.00
16	BUTTONS PER BUTTON-DOWN SHIRT	12.00
17	TOTAL BUTTONS AVAILABLE	110000000

FIGURE D-26 Shirt manufacturing constants

- TAX RATE: The rate is .28 on pretax income, but no taxes are paid on losses.
- SELLING PRICE: You sell polo-style T-shirts for $8 and button-down shirts for $36.
- VARIABLE COST TO MAKE: It costs $2.50 to make a T-shirt and $14 to make a button-down shirt. These variable costs are for machine operator labor, cloth, buttons, etc.
- COTTON USAGE: Each polo T-shirt uses 1.5 pounds of cotton fabric. Each button-down shirt uses 2.5 pounds of cotton fabric.
- TOTAL COTTON AVAILABLE: You have only 13 million pounds of cotton on hand to make all of the T-shirts and button-down shirts.
- BUTTONS: Each polo T-shirt has 3 buttons. By contrast, each button-down shirt has 1 button on each collar tip, 8 buttons down the front, and 1 button on each cuff, for a total of 12 buttons. You have 110 million buttons on hand to be used to make all of your shirts.

CALCULATIONS Section

Your spreadsheet should contain the calculations shown in Figure D-27.

	A	B
19	CALCULATIONS	
20	RATIO OF NET INCOME TO TOTAL REVENUE	
21	COTTON USED: T-SHIRTS	
22	COTTON USED: BUTTON-DOWN SHIRTS	
23	COTTON USED: TOTAL	
24	BUTTONS USED: T-SHIRTS	
25	BUTTONS USED: BUTTON-DOWN SHIRTS	
26	BUTTONS USED: TOTAL	
27	RATIO OF BUTTON-DOWNS TO T-SHIRTS	

FIGURE D-27 Shirt manufacturing calculations

Calculations (and related business constraints) are discussed next.

- RATIO OF NET INCOME TO TOTAL REVENUE: The minimum return on sales (net income after taxes divided by total revenue) is .20.
- COTTON USED/BUTTONS USED: You have a limited amount of cotton and buttons. The usage of each resource must be calculated, then used in constraints.
- RATIO OF BUTTON-DOWNS TO T-SHIRTS: You think you must make at least 2 million T-shirts and at least 2 million button-down shirts. You want to be known as a balanced shirtmaker, so you think the ratio of button-downs to T-shirts should be no greater than 4:1. (Thus, if 9 million button-down shirts and 2 million T-shirts were produced, the ratio would be too high.)

INCOME STATEMENT Section

Your spreadsheet should have the income statement skeleton shown in Figure D-28.

	A	B
29	INCOME STATEMENT	
30	T-SHIRT REVENUE	
31	BUTTON-DOWN SHIRT REVENUE	
32	TOTAL REVENUE	
33	VARIABLE COSTS: T-SHIRTS	
34	VARIABLE COSTS: BUTTON-DOWNS	
35	TOTAL COSTS	
36	INCOME BEFORE TAXES	
37	INCOME TAX EXPENSE	
38	NET INCOME AFTER TAXES	

FIGURE D-28 Shirt manufacturing income statement line items

The Solver's target is net income, which must be maximized.

Use the table shown in Figure D-29 to write out your constraints before entering them into the Solver.

Expression in English	Fill in the Excel Expression
Net income to revenue	___ >= ___
Ratio of BDs to Ts	___ <= ___
Min T-shirts	___ >= ___
Min button-downs	___ >= ___
Usage of buttons	___ <= ___
Usage of cotton	___ <= ___

FIGURE D-29 Logic of shirt manufacturing constraints

When you are finished with the program, print the sheets. Then use the Office button to save the file, close it, and choose Exit to leave Excel.

TROUBLESHOOTING THE SOLVER

Use this section to overcome problems with the Solver and to review some Windows file-handling procedures.

Rerunning a Solver Model

Assume you have changed your spreadsheet in some way and want to rerun the Solver to get a new set of answers. (For example, you may have changed a constraint or a formula in your spreadsheet.) Before you click Solve to rerun the Solver, put the number 1 in the changing cells.

Creating Overconstrained Models

It is possible to set up a model that has no logical solution. For example, in the second version of the sporting goods problem, suppose you had specified that at least 1 million basketballs were needed. When you clicked Solve, the Solver would have tried to compute an answer, but then would have admitted defeat by telling you that no feasible solution was possible, as shown in Figure D-30.

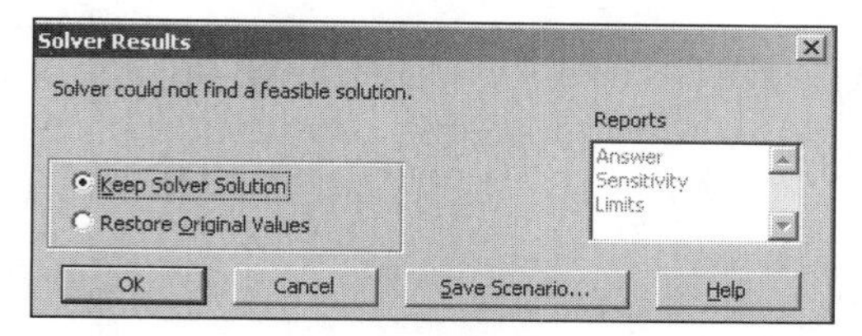

FIGURE D-30 Solver Results message: Solution not feasible

In the Reports window, the choices (Answer, Sensitivity, and Limits) would be in gray, indicating that they are not available as options. Such a model is sometimes referred to as overconstrained.

Setting a Constraint to a Single Amount

You may want an amount to be a specific number, as opposed to a number in a range. For example, if the number of basketballs needed to be exactly 30,000, then the "equals" operator would be selected, as shown in Figure D-31.

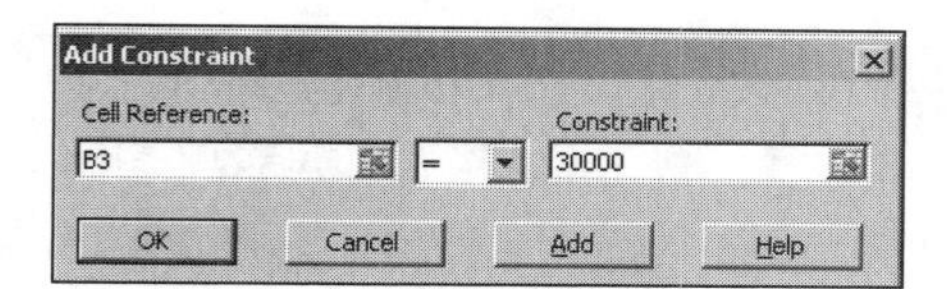

FIGURE D-31 Constraining a value to equal a specific amount

Setting a Changing Cell to an Integer

You may want to force changing cell values to be integers. If so, select the Int operator in the Add Constraint window, as described in a prior section.

Forcing the Solver to find only integer solutions slows down the Solver. In some cases, the change in speed can be noticeable to the user. It also can prevent the Solver from seeing a feasible solution when one could be found using noninteger answers. For those reasons, it's usually best not to impose the integer constraint unless the logic of the problem demands it.

Deleting Extra Answer Sheets

Suppose you've run different scenarios, each time asking for an Answer Report. As a result, you have a number of Answer Report sheets in your Excel file, but you don't want to keep them all. How do you get rid of an Answer Report sheet? Follow this procedure: First, select the Home tab. In the Cells group, click the Delete drop-down arrow and select Delete Sheet. You will *not* be asked if you really mean to delete. Therefore, make sure you want to delete the sheet before you act.

Restarting the Solver with All New Constraints

Suppose you wanted to start over with a new set of constraints. In the Solver Parameters window, click Reset All. You will be asked if you really mean to do that, as shown in Figure D-32.

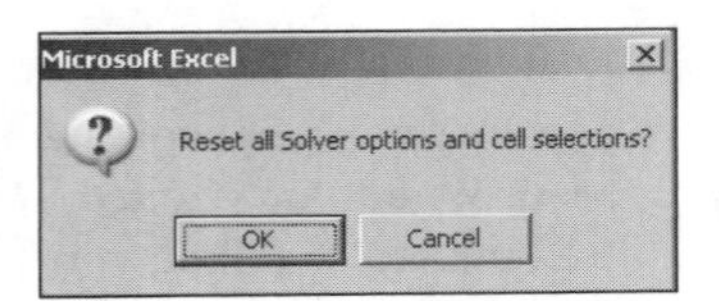

FIGURE D-32 Reset options warning query

If you do, select OK. That gives you a clean slate, with all entries deleted, as shown in Figure D-33.

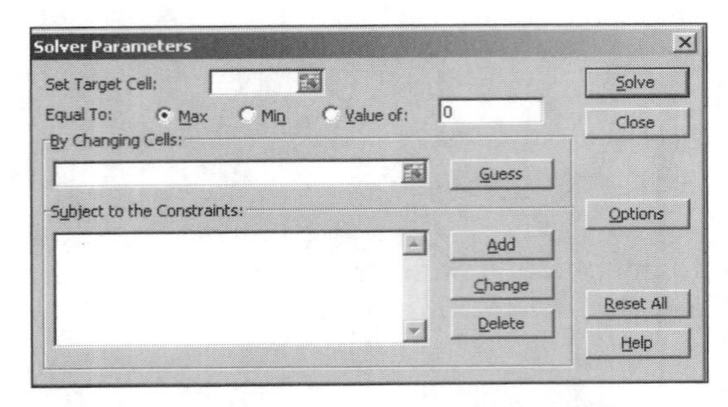

FIGURE D-33 Reset Solver Parameters window

As you can see, the target cell, changing cells, and constraints have been reset. From this point, you can specify a new model.

> **NOTE**
>
> If you select Reset All, you really are starting over. If you merely want to add, delete, or edit a constraint, do not use Reset All. Use the Add, Delete, or Change buttons as appropriate.

Using the Solver Options Window

The Solver has a number of internal settings that govern its search for an optimal answer. If you click the Options button in the Solver Parameters window, you will see the defaults for those settings, as shown in Figure D-34.

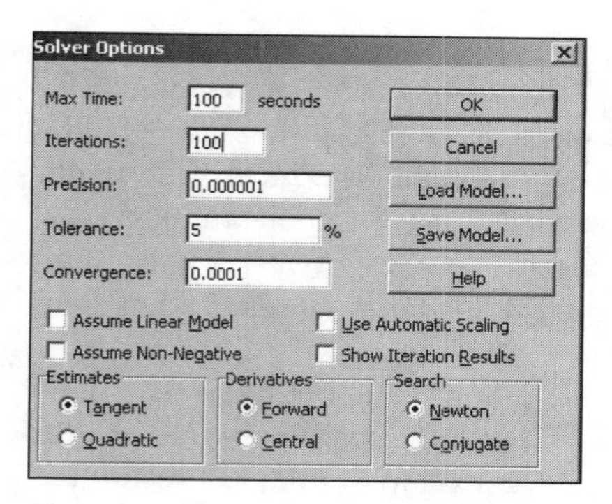

FIGURE D-34 Solver Options window with default settings for Solver parameters

In general, Solver Options govern how long the Solver works on a problem and/or how precise it must be in satisfying constraints. You should not check Assume Linear Model if changing cells are multiplied or divided (as they are in this book's cases) or if some of the spreadsheet's formulas use exponents.

You should not have to change these default settings for the cases in this book. If you think that your Solver work is correct but the Solver cannot find a feasible solution, you should check to see that Solver Options are set as shown in Figure D-34.

Printing Cell Formulas in Excel

To show the Cell Formulas on the screen, press Ctrl and the left quote (‘) keys at the same time: Ctrl-‘. The left quote is usually on the same key as the tilde (~). Pressing Ctrl-‘ automatically widens cells so the formulas can be read. You can change cell widths by clicking and dragging at the column indicator (A, B, C, etc.) boundaries.

To print the formulas, use the Office button and select Print. Print the sheet as you would normally. To restore the screen to its typical appearance (showing values, not formulas), press Ctrl-‘ again. (It's like a toggle switch.) If you did not change any column widths while in the cell formula view, the widths will be as they were.

Reviewing the Printing, Saving, and Exiting Procedures

Print the Solver spreadsheets in the normal way. Activate the sheet, then select the Office button and Print. You can print an Answer Report sheet the same way.

To save a file, use the Office button and select Save (or Save As). Make sure you select the proper drive (for example, drive A:) if you intend for your file to be on a disk. When exiting from Excel, always start with the Office button. Then select Close (with the disk in drive A: if your file is on the disk) and select Exit. Only then should you take the disk out of drive A:.

NOTE

If you use File—Exit (without selecting Close first), you risk losing your work.

Sometimes the Solver will come up with strange results. For instance, your results might differ from the target answers that your instructor provides for a case. Thinking that you've done something wrong, you ask to compare your cell formulas and constraint expressions with those your instructor created. Lo and behold, you see no differences! Surprisingly, and for no apparent reason, the Solver occasionally produces slightly different outputs from inputs that are seemingly the same. This may occur because, for your application, the order of the constraints matters, or even the order in which they are entered matters. In any case, if you are close to the target answers but cannot see any errors, it's best to see your instructor for guidance, rather than to spin your wheels.

As another example, assume you ask for integer changing cell outputs. The Solver may tell you that the correct output is 8.0000001, or 7.9999999. In such situations, the Solver is apparently not sure about its own rounding. In that case, you should merely humor the Solver and (continuing the example) take the result as the integer 8, which is what the Solver was trying to report in the first place.

CASE 8

THE NEW FUND MANAGER INVESTMENT MIX DECISION

Decision Support Using Excel Solver

PREVIEW

You are a new investment manager working for a giant mutual fund. You will have $50 million to invest. You want to buy securities that are profitable but safe. In this case, you will use Excel to determine the best mix of securities to buy for your fund.

PREPARATION

- Review spreadsheet concepts discussed in class and in your textbook.
- Complete any exercises that your instructor assigns.
- Complete any part of Tutorial D that your instructor assigns, or refer to Tutorial D as necessary.
- Review file-saving procedures for Windows programs, as discussed in Tutorial C.
- Refer to Tutorials E and F as necessary.

BACKGROUND

You work for a gigantic mutual fund that manages billions of dollars. The money is parceled out to its investment managers to be invested. Managers who invest well are retained by the company. You have recently been appointed as an investment manager, and you will have $50 million to invest for the year. You want to consider the following kinds of investments:

- Bonds issued by the U.S. government—These bonds earn interest and are considered very safe, because the government has a printing press and can always repay its debts.
- AAA corporate bonds—These bonds earn interest. The credit worthiness of corporations is rated. "AAA" is the highest rating in the markets, so these bonds are considered very safe.
- Blue chip common stocks—Shares of common stock represent ownership in corporations. "Blue chip" corporations such as General Electric are considered the most successful companies.
- Cash—Cash invested in money market instruments does not earn much interest, but holding cash is considered extremely safe.

Rates of Return and Risk

Each kind of security has an expected "rate of return" and an expected level of risk. Assume that the rate of return is one of the following:

- The amount of cash received by the investor (in dividends or interest) divided by the cost of the investment
- The increase in the security's market price divided by the cost of the investment

Investors who seek higher rates of return must assume higher levels of risk. When the market value of a stock or bond falls after being purchased, the decrease is called "a loss of principal." For example, suppose that an investor buys 1,000 shares of ABC stock at $80 dollars per share. If ABC shares then fall to $10 per share, the loss of principal is 1,000 times $70, or a loss of $70,000. You understand that an investor who faces a higher level of risk will demand a higher rate of return.

Your fund's Research Department has estimated the expected annual rate of return for the three kinds of securities and for cash. The rates are shown in Figure 8-1.

Type of Investment	Expected Rate of Return
U.S. government bonds	.0350
AAA corporate bonds	.0450
Blue-chip stocks	.0475
Cash	.0050

FIGURE 8-1 Expected rates of return

For example, suppose that $1 million is invested in U.S. government bonds. During the year, 3.5% of that amount—$35,000—will be received on that investment. That $35,000 will be "revenue" in the income statement. As another example, suppose that $1 million is invested in blue-chip stocks. If the prices of the stocks go up an average of 4.75% per year, the increase will yield $47,500 of revenue in the income statement. Notice that the rates of return cited here are annual rates.

The fund's Research Department has investment managers assigned to monitor the expected levels of risk for different kinds of securities. Risk is denoted by a numerical Expected Risk Indicator. The greater this number is, the higher the risk of losses. The expected risk indicators are shown in Figure 8-2.

Type of Investment	Expected Risk Indicator
U.S. government bonds	1
AAA corporate bonds	2
Blue-chip stocks	3
Cash	0

FIGURE 8-2 Expected risk indicators

As shown in Figure 8-2, U.S. government bonds have very low risk because the government can be expected to repay its debts (though the market value of any bond can decline). AAA corporate bonds have some risk, but not much, because the best companies almost always pay their debts. Blue-chip stocks carry some risk—the risk on stock prices is generally higher than on bond prices. There is almost no risk of losing cash that is invested in money-market instruments, so zero risk is assumed.

Expenses

Operating the mutual fund generates two kinds of expenses. First, the fund must work with a separate management company that keeps financial records on computers, provides frequent data and reports from Wall Street, and so on. Your fund's management company charges three-quarters of one percent of the amount invested in non-cash securities. For example, if you invested $49 million in non-cash securities, you would incur a management fee of 0.0075 * $49,000,000.

Second, you and your analysts must watch each kind of security. The cost of this activity (salaries, office expenses, and so on) depends on the kind of security. Assume that high-risk securities require more oversight, so the expenses associated with a high-risk security are greater than those for a low-risk security. The fund's Research Department has estimated the expense factors for securities and cash (see Figure 8-3).

Type of Investment	Expense Factor
U.S. government bonds	.0015
AAA corporate bonds	.0035
Blue-chip stocks	.0055
Cash	.0005

FIGURE 8-3 Investment expense factors

For example, suppose that $1 million is invested in U.S. government bonds. During the year, 0.15% of that sum—$1,500—will be spent monitoring the $1 million. Thus, the $1,500 will be a "variable" expense in your income statement for the year.

Your manager has given you a performance goal: The net income to total investment ratio must be at least 2.5% (.025). That is, the fund's net income after taxes, divided by the amount of dollars invested, must be at least .025. For example, if yearly net income after taxes is $1 million on $10 million invested, then the net income to total investment would be 10%, and your manager would be satisfied. By contrast, if yearly net income after taxes is $100,000 on the $10 million invested, your manager would not be satisfied with the 1% ratio.

Investment Requirements

You will have $50 million to invest for the year, and you must invest it all. Your manager says that you must put at least $250,000 but no more than $5 million into cash. You must invest at least $5 million in U.S. government bonds, at least $5 million in AAA corporate bonds, and at least $5 million in blue-chip stocks. Your non-cash investments must total at least $45 million. The ratio of dollars invested in U.S. government bonds and AAA corporate bonds to the total non-cash investment must be at least 0.50. For example, if $15 million is invested in each of U.S. government bonds, AAA corporate bonds, and blue-chip stocks, the ratio would be 30/45 = .667.

Your manager has set a quantifiable risk-level boundary. The "weighted average risk level" on the money invested each month must be at least 1.5, but should not exceed 3.5. The weighted average risk level is computed by the following procedure:

- weight the expected risk indicators by the amount invested in each kind of security
- add the weighted amounts
- divide the total by the total amount invested

To use a simplified example, suppose that only U.S. government bonds and blue-chip stocks are purchased. Suppose that $5 million is invested in the bonds and $5 million in the stocks. The weighted average risk level would be 2.0:

$$((\$5{,}000{,}000 * 1) + (\$5{,}000{,}000 * 3))/\$10{,}000{,}000 = 2.0$$

(This example only considers two kinds of securities, but you will actually invest in more than two kinds.)

Your manager thinks that the scenario outlined here is a proper plan for you in the near future. She calls this scenario the "base case." However, you should know that executives in the fund's senior management think that fund managers should be more aggressive, and this attitude may change the rules in the future.

Your manager wants you to set up a "base case" plan for your $50 million of investments. The goal should be to maximize your investments' net income after taxes without assuming too much risk. Your manager knows that you can model this problem in the Solver because you said you could when you were interviewed for the job!

ASSIGNMENT 1: CREATING A SPREADSHEET FOR DECISION SUPPORT

In this assignment, you will produce spreadsheets that model the business decision. In Assignment 1A, you will create a Solver spreadsheet to model the investment decision. This model will be the base case. In Assignment 1B, the extension case, you will create a second Solver spreadsheet to model the investment decision given more aggressive operating rules.

In Assignments 2 and 3, you will use the spreadsheet models to develop information needed to recommend the best mix for your investments. In Assignment 2B, you will document your recommendations in a memorandum. In Assignment 3, you will give your recommendations in an oral presentation.

Your spreadsheets for this assignment should include the following sections. You will be shown how to set up each section before entering cell formulas. Your spreadsheets will also include decision constraints that you will enter using the Solver.

- Changing Cells
- Constants
- Calculations
- Income Statement

Assignment 1A: Creating the Spreadsheet—Base Case

A discussion of each spreadsheet section follows. You will learn how to set up each section and learn the logic of the formulas in the sections' cells. When you enter data in the spreadsheet skeleton, follow the order shown in this section. *The spreadsheet skeleton is available for you to use; you can choose to type in it or not.* To access the base case spreadsheet skeleton, go to your data files, select Case 8, and then select **FundManager.xlsx**.

Changing Cells Section

Your spreadsheet should include the changing cells shown in Figure 8-4.

	A	B	C
1	FUND MANAGER INVESTMENT MIX PROBLEM		
2			BASE CASE
3	CHANGING CELLS		
4	DOLLARS INVESTED IN U.S. BONDS		$1
5	DOLLARS INVESTED IN AAA CORPORATE BONDS		$1
6	DOLLARS INVESTED IN BLUE CHIP STOCKS		$1
7	DOLLARS INVESTED IN CASH		$1

FIGURE 8-4 Changing Cells section

You will ask the Solver model to compute how many dollars to invest in each kind of security during the year. Start with a "1" in each cell. The Solver will change each 1 as it computes the answer. The Solver can recommend investing a fractional part of a dollar for a security.

Constants Section

Your spreadsheet should include the constants shown in Figure 8-5. An explanation of the line items follows the figure.

	A	B	C
10	CONSTANTS		
11	TAX RATE		0.20
12	RATE OF RETURN:	---	
13	U.S. BONDS		0.0350
14	AAA CORPORATE BONDS		0.0450
15	BLUE CHIP STOCKS		0.0475
16	CASH		0.0050
17			
18			
19	MANAGEMENT FEE %		0.0075
20	RISK LEVEL ASSIGNED:	---	
21	U.S. BONDS		1
22	AAA CORPORATE BONDS		2
23	BLUE CHIP STOCKS		3
24	CASH		0
25			
26			
27	EXPENSE FACTOR:	---	
28	U.S. BONDS		0.0015
29	AAA CORPORATE BONDS		0.0035
30	BLUE CHIP STOCKS		0.0055
31	CASH		0.0005

FIGURE 8-5 Constants section

- Tax Rate—The tax rate is applied to income before taxes.
- Rate of Return—The rate of return for each type of security. The rate of return is used to compute the revenue from each security.
- Management Fee—A flat percentage of the total amount invested, except that the fee is not charged on cash invested. (There is nothing to "manage" in that case.)
- Risk Level Assigned—The risk indicator for each type of security.
- Expense Factor—This percentage is used to compute the expense associated with each investment.

Calculations Section

Your spreadsheet should calculate the amounts shown in Figure 8-6. These amounts will be used in the Income Statement section and the Constraints section. Calculated values may be based on the values of the changing cells, the constants, and other calculations. An explanation of the line items follows the figure.

	A	B	C
34	CALCULATIONS		
35	REVENUE:		---
36	U.S. BONDS		
37	AAA CORPORATE BONDS		
38	BLUE CHIP STOCKS		
39	CASH		
40			
41			
42	VARIABLE EXPENSES:		---
43	U.S. BONDS		
44	AAA CORPORATE BONDS		
45	BLUE CHIP STOCKS		
46	CASH		
47			
48			
49	WEIGHTED AVG RISK FACTOR		
50	TOTAL NON-CASH INVESTED		
51	TOTAL INVESTED		
52	TOTAL U.S. GOVERNMENT AND AAA		
53	PERCENT IN U.S. GOVERNMENT AND AAA		
54	NET INCOME TO TOTAL INVESTED RATIO		

FIGURE 8-6 Calculations section

- Revenue—Compute the revenue for each security. Revenue is a function of the amount invested in the security and the security's annual rate of return. The amount invested is shown in the Changing Cells section. The rate of return is taken from the Constants section.
- Variable Expenses—Compute each security's variable expense, which is a function of the amount invested and the security's expense ratio. The expense ratio is taken from the Constants section.
- Weighted Avg Risk Factor—Compute the weighted average risk factor for the four investment types. Follow the logic of the example explained earlier.
- Total Non-Cash Invested—Compute the total dollars invested in non-cash securities.
- Total Invested—Compute the total invested in all securities (including cash).
- Total U.S. Government and AAA—Compute the total invested in U.S. government bonds and AAA corporate bonds.
- Percent in U.S. Government and AAA—Compute the ratio of U.S. government bonds and AAA corporate bonds to total non-cash dollars invested.
- Net Income to Total Invested Ratio—Compute the ratio of net income after taxes to total dollars invested in all securities.

Income Statement Section

The statement shown in Figure 8-7 is the projected net income for a year on the amount invested. An explanation of the line items follows the figure.

	A	B	C
57	**INCOME STATEMENT**		
58	TOTAL REVENUE		
59	EXPENSES:		---
60	MANAGEMENT FEE		
61	VARIABLE EXPENSES		
62	TOTAL EXPENSES		
63	INCOME BEFORE TAXES		
64	INCOME TAX EXPENSE		
65	NET INCOME AFTER TAXES		

FIGURE 8-7 Income Statement section

- Total Revenue—This amount is computed by totaling the previously calculated revenues.
- Management Fee—A function of the management fee expense rate (from the Constants section) and non-cash amounts invested.
- Variable Expenses—This amount is computed by totaling the previously calculated variable expenses.
- Total Expenses—The sum of the management fee and variable expenses.
- Income Before Taxes—Total revenue minus total expenses.
- Income Tax Expense—This expense is zero if income before taxes is zero or less; otherwise, apply the tax rate to income before taxes.
- Net Income After Taxes—This amount is income before taxes minus income taxes.

Constraints and Running the Solver

In this part of the assignment, you determine the decision constraints. Enter the base case decision constraints using the Solver. Run the Solver and ask for the Answer Report when the Solver has found a solution that satisfies the constraints.

When you finish, print the entire workbook, including the Solver Answer Report sheet. Save the workbook by opening the File menu and selecting Save; FundManager.xlsx is suggested as the filename. Then, to prepare for the extension case, open the File menu and select Save As to create a new spreadsheet. (**FundManager2.xlsx** would be a good filename.)

Assignment 1B: Creating the Spreadsheet—Extension Case

Some senior managers want investment managers to invest in Credit Default Swaps (CDS) because they are seen as a low-risk and relatively inexpensive way to increase the bottom line.

If you do not know what a credit default swap is, consider the following example. Assume that a large pension fund invests in $1 million of XYZ corporation's bonds. The fund manager, for some reason, is concerned that XYZ will not pay off the bonds when they come due. In other words, the manager is worried that XYZ will "default" and that bondholders (creditors) will then lose their investment. The pension manager is willing to pay a fee to insure against losing $1 million. He looks for someone to agree (for a fee) to pay the value of the $1 million bonds if XYZ defaults. Assume that the KLM company agrees to provide this insurance. If XYZ does not default, KLM pockets the fee and has no expense. But if XYZ defaults, KLM must pay $1 million to the pension fund.

In reality, corporations rarely default on their bonds, so CDS fees are easy money for the insurer. In other words, the insurer rarely has to pay the value of the bonds it insured.

Wall Street has an active CDS market, and many huge pension funds want to have their investments insured. Your fund's senior management wants to get involved as a CDS insurer so the company can collect the fees. Senior managers say that only AAA debt would be insured and that they are not afraid of the default risk.

In your changing cells, you specify the total dollar value of credit that you will insure. The changing cells would look like Figure 8-8.

	A	B	C
1	**FUND MANAGER INVESTMENT MIX PROBLEM**		
2			**EXTENSION CASE**
3	**CHANGING CELLS**		
4	DOLLARS IN U.S. BONDS		$1
5	DOLLARS IN AAA CORPORATE BONDS		$1
6	DOLLARS IN BLUE CHIP STOCKS		$1
7	DOLLARS IN CASH		$1
8	PENSION DOLLARS INSURED BY CDS		$1

FIGURE 8-8 Extension case Changing Cells section

Assume that management thinks you should insure at least $50 million but not more than $100 million. The rate of return on CDS would be 3%, so CDS revenue would be 3% of the amount insured. (The spreadsheet should have a blank row for this factor.) Assume that you would not monitor the status of the bonds insured, so the variable expense rate for CDS would be zero.

Note that you would not be "investing" in the bonds insured; you would merely insure against their eventual default. Thus, the total invested would remain at $50 million.

Although management thinks the risk of default is low, they think the insured bonds should be included in the calculation of the weighted average risk factor. The same risk level used for corporate bonds should be used for bonds insured. (The spreadsheet should have a blank row for this risk level.) The weighted average risk should stay between 1.5 and 3.5 in the extension case.

The same minimum investment levels would prevail in this extension case, and the management fee would be computed in the same way. In this more aggressive model, management does not think you should worry about the relationship of U.S. government and AAA bonds to total investment.

Again, the goal is to maximize net income. The ratio of net income to total investment must be at least 2.5%, as it was in the base case. Management thinks that achieving this goal will be easier in the extension case; after all, revenue should be augmented by the easy CDS money!

Modify the extension case spreadsheet to handle the more aggressive scenario. Modify the constraints as needed. Run the Solver and ask for the Answer Report when the Solver finds a solution that satisfies the constraints.

When you finish, print the entire workbook, including the Solver Answer Report sheet. Save the workbook, close the file, and exit Excel.

ASSIGNMENT 2: USING THE SPREADSHEET FOR DECISION SUPPORT

You have built the base case and extension case models because you want to know the investment mix for each scenario and which scenario yields the highest net income after taxes, consistent with perceived risks. You will now complete the case by using the Answer Reports to gather the data needed to make the investment mix decisions and by documenting your recommendations in a memorandum.

Assignment 2A: Using the Spreadsheet to Gather Data

You have printed the Answer Report sheets for each scenario. Each sheet reports the dollar amount of each kind of security to purchase in a month, plus the target net income and risk in each case. Summarize key data in a table in your memorandum. The table format is shown in Figure 8-9.

	Base Case	Extension Case	Difference
Net income after taxes (in dollars)			
Net income to total investment ratio			
Weighted average risk ratio			
Dollar value of bonds insured by CDS	$0		

FIGURE 8-9 Format of table to insert in memorandum

Assignment 2B: Documenting Your Recommendation in a Memorandum

Use Microsoft Word to write a brief memorandum to your manager. State the results of your analysis and recommend which investment strategy to adopt: the base case or the extension case. Keep in mind that your management ranks investment managers by their net income ratio and that a default on $50 million in bonds, while unlikely, would completely wipe out the value of your investment portfolio. Is the easy CDS money worth the apparently small risk? In your memo, observe the following requirements:

- Set up your memorandum as discussed in Tutorial E.
- In the first paragraph, briefly define the situation and state the purpose of your analysis.
- Then describe your results and state your recommendations. State the logic that supports your recommendations.
- Support your statements graphically by including the summary table shown in Figure 8-9. The procedure for creating a table in Word is described in Tutorial E.

ASSIGNMENT 3: GIVING AN ORAL PRESENTATION

Your instructor may request that you present your analysis and recommendations in an oral presentation. If so, assume that your manager wants you to explain your analysis and recommendations to her and other investment managers in 10 minutes or less. Use visual aids or handouts that you think are appropriate. Tutorial F explains how to prepare and give an oral presentation.

DELIVERABLES

Assemble the following deliverables for your instructor:

1. A printout of your memorandum
2. Spreadsheet printouts
3. Electronic media such as a USB key or CD, which should include your Word memo and your Excel spreadsheet file

Staple the printouts together with the memorandum on top. If you have more than one .xlsx file on your electronic media, write your instructor a note that identifies the name of your model's .xlsx file.

CASE 9

THE GOLF CLUB PRODUCT MIX DECISION

Decision Support Using Excel Solver

PREVIEW

You have started a golf club manufacturing company that will make sets of clubs for men and women. In this case, you will use Excel to determine how many sets of men's and women's clubs to produce.

PREPARATION

- Review spreadsheet concepts discussed in class and in your textbook.
- Complete any exercises that your instructor assigns.
- Complete any part of Tutorial D that your instructor assigns, or refer to Tutorial D as necessary.
- Review file-saving procedures for Windows programs, as discussed in Tutorial C.
- Refer to Tutorials E and F as necessary.

BACKGROUND

You worked for a large golf club manufacturer for a decade. You developed your own ideas for how clubs should be made, so you have started your own golf club manufacturing company. Your company will focus on regional sales. You plan to make a high-quality product that sells for less than the national brands.

You think that clubs should be sold in sets, not individually. A set of your clubs has two "woods" and eight "irons." Your sets do not include a putter, the club used when the ball is on the green near the hole. Sets of clubs are designed for men and for women. Men's clubs are typically longer and heavier than women's clubs, and have some other differences.

A golf club has two main parts called a head and a shaft, which are made separately and then assembled to make the finished club.

Woods are used for long shots: for the "drive" from the tee and for shots from the fairway. In the old days, the head was actually made of wood, but a modern wood's head is made from hollow metal. (You will use stainless steel.)

Irons are typically used for shots into the green from 180 yards or less. The heads of irons were once made from actual iron, but a modern iron's head is made from different metals. (Again, you will use stainless steel.) Your clubs will include eight irons, including a pitching wedge, a sand wedge, and irons numbered 4 through 9.

In the past, club shafts were made from wood, but today a variety of materials are used, including steel, titanium, and graphite. The end of the shaft that the golfer holds has a grip, which is generally made of rubber or leather. You will use graphite and rubber, respectively.

Each club has a different loft, which is the angle that the face makes with the ground. Woods have little loft to allow the golfer to hit the ball low and long. Your driver will have 10 degrees of loft and your 3-wood will have 15 degrees of loft. A 4-iron has 23 degrees of loft, and its shaft is almost as long as a wood's. A typical male golfer can hit a 4-iron shot 170 yards. Your 9-iron has 41 degrees of loft and a short shaft; a golfer typically hits a 9-iron shot higher in the air than a 4-iron and not as far, but probably with greater accuracy.

The sand wedge has 55 degrees of loft and a heavy head; it is used to blast the ball from green-side sand traps. A pitching wedge has 45 degrees of loft and is used for precision shots close to the green.

Your clubs will be manufactured in a three-step process:

1. Investment casting
2. Tube drawing
3. Assembly

Each of these steps is described next.

Investment Casting

You have a master die of the club's head. Molten metal is poured into the die and allowed to harden. Woods are actually hollow and more complex to make than irons. More time is needed to cast a wood's head than an iron's head.

Tube Drawing

A metal tube of the desired length is heated and drawn through a die to give the tube a tapered shape—widest at the grip end and narrowest where the shaft meets the club head. The process is repeated for each shaft until it has the desired tapering.

Assembly

The narrow part of the shaft is inserted into the head and bonded with a strong adhesive so that the face of the club head lies "square" when on the ground. That is, the face lies perpendicular to the ball's intended line of flight. If the angle is not correct, shots will go awry!

The grip is then molded onto the other end of the shaft. Finally, the club is scanned and X-rayed for quality control.

You have acquired the machinery needed for each manufacturing step. The processing times required at each step are shown in Figure 9-1.

	Investment Casting	Tube Drawing	Assembly
Set of men's clubs	42	54	54
Set of women's clubs	38	46	54

FIGURE 9-1 Processing times (in minutes) for sets of clubs

Approximately 42 minutes of machine processing time are required to cast the 10 clubs in a men's set. Women's club heads are smaller than men's. Only 38 minutes of machine processing time are required to cast the 10 clubs in a women's set. The shafts for men's clubs are longer than women's, so tube drawing for men's clubs requires more processing time than women's clubs. The assembly times are the same for each kind of club set.

The amounts of machine processing times available in a year are shown in Figure 9-2.

	Investment Casting	Tube Drawing	Assembly
Available capacity (minutes)	100,000	105,000	110,000

FIGURE 9-2 Processing capacity available (in minutes)

You have established these times based on your experience and on test runs in your plant. Thus, 100,000 minutes are available in the year for investment casting of all men's and women's clubs produced. A total of 105,000 minutes are available for tube drawing of the clubs, and 110,000 minutes are available for assembly.

You may wonder why capacity differs at each step. The differences are due to machine servicing down time. For example, investment casting machinery is harder to clean and maintain than other machinery, so it is offline more often.

You have carefully calculated the cost of materials, labor, and overhead for men's and women's clubs. You think that direct production costs will be $900 for a men's set and $885 for a women's set. You plan to sell a men's set for $1,200 and a women's set for $1,140. Selling, general, and administrative (SG&A) expenses are expected to be $100,000 in the year; these SGA expenses are not directly associated with any one product.

To ensure that you do not run out of a product, you will make at least 700 sets of men's clubs and at least 700 sets of women's clubs. Of course, you will not make a partial set of clubs.

You think you can sell all the sets you make, and you want to maximize net income after taxes. Therefore, how many sets of men's clubs and women's clubs should you make? The answer to that question is the focus of your "base case."

As a separate issue, one of the big manufacturers has come to you with an offer. The manufacturer sells its popular "Earthquake" putter for $500 apiece. (It is amazing how much golfers will pay for a club that they think will improve their score!) The manufacturer produces many more putter heads and shafts than it can sell. You could take the excess, then make and sell an Earthquake "knock-off" putter as your own product for much less than $500. Perhaps adding a putter to your products would yield a better bottom line. You call this option the "extension case."

You are confident that you can model the base case and extension case problems in the Solver.

ASSIGNMENT 1: CREATING A SPREADSHEET FOR DECISION SUPPORT

In this assignment, you will produce spreadsheets that model the business decision. In Assignment 1A, you will create a Solver spreadsheet to model the base case production decision. In Assignment 1B, you will create a second Solver spreadsheet to model the extension case production decision.

In Assignments 2 and 3, you will use the spreadsheet models to develop information needed to recommend the best production mix. In Assignment 2B, you will document your recommendations in a memorandum. In Assignment 3, you will give your recommendations in an oral presentation.

Your spreadsheets for this assignment should include the following sections. You will be shown how to set up each section before entering cell formulas. Your spreadsheets will also include decision constraints that you will enter using the Solver.

- Changing Cells
- Constants
- Calculations
- Income Statement

Assignment 1A: Creating the Spreadsheet—Base Case

A discussion of each spreadsheet section follows. You will learn how to set up each section and learn the logic of the formulas in the sections' cells. When you enter data in the spreadsheet skeleton, follow the order shown in this section. *The spreadsheet skeleton is available for you to use; you can choose to type in it or not.* To access the base case spreadsheet skeleton, go to your data files, select Case 9, and then select **GolfSets.xlsx**.

Changing Cells Section

Your spreadsheet should include the changing cells shown in Figure 9-3.

	A	B	C
1	**GOLF CLUB PRODUCT MIX PROBLEM**		
2			**BASE CASE**
3	**CHANGING CELLS**		
4	NUMBER OF SETS OF MEN'S CLUBS		1
5	NUMBER OF SETS OF WOMEN'S CLUBS		1

FIGURE 9-3 Changing Cells section

You will ask the Solver model to compute how many sets of men's and women's clubs to produce. Start with a "1" in each cell. The Solver will change each 1 as it computes the answer. The Solver cannot recommend building a fractional part of a golf club set.

Constants Section

Your spreadsheet should include the constants shown in Figure 9-4. An explanation of the line items follows the figure.

	A	B	C
9	**CONSTANTS**		
10	TAX RATE		0.28
11	SELLING PRICES:		---
12	MEN'S SET		$1,200
13	WOMEN'S SET		$1,140
14			
15			
16	PRODUCTION COST:		---
17	MEN'S SET		$900
18	WOMEN'S SET		$885
19			
20			
21	TIME REQUIREMENTS (MINUTES):		---
22	INVESTMENT CASTING, MEN'S SET		42
23	INVESTMENT CASTING, WOMEN'S SET		38
24	TUBE DRAWING, MEN'S SET		54
25	TUBE DRAWING, WOMEN'S SET		46
26	ASSEMBLY, MEN'S SET		54
27	ASSEMBLY, WOMEN'S SET		54
28			
29			
30	DEPT TIME AVAILABLE IN YEAR (MINUTES):		---
31	INVESTMENT CASTING		100,000
32	TUBE DRAWING		105,000
33	ASSEMBLY		110,000
34			
35	SELLING, GENERAL & ADMINISTRATIVE		$100,000

FIGURE 9-4 Constants section

- Tax Rate—The tax rate is applied to income before taxes.
- Selling Price, Men's Set—You will sell a set of men's clubs for $1,200.
- Selling Price, Women's Set—You will sell a set of women's clubs for $1,140.
- Production Cost, Men's Set—The direct production cost for a men's set will be $900.
- Production Cost, Women's Set—The direct production cost for a women's set will be $885.
- Time Requirements (minutes)—The times required at each production stage for each kind of set.
- Dept Time Available in Year (minutes)—The times available at each production stage.
- Selling, General & Administrative—SG&A expenses are expected to be $100,000 in the year.

Calculations Section

Your spreadsheet should calculate the amounts shown in Figure 9-5. These amounts will be used in the Income Statement section and the Constraints section. Calculated values may be based on the values of the changing cells, the constants, and other calculations. An explanation of the line items follows the figure.

	A	B	C
37	CALCULATIONS		
38	REVENUE:		---
39	MEN'S SETS		
40	WOMEN'S SETS		
41			
42	PRODUCTION COSTS:		---
43	MEN'S SETS		
44	WOMEN'S SETS		
45			
46	TIME USED (MINUTES):		---
47	INVESTMENT CASTING, MEN'S + WOMEN'S SETS		
48	TUBE DRAWING, MEN'S + WOMEN'S SETS		
49	ASSEMBLY, MEN'S + WOMEN'S SETS		

FIGURE 9-5 Calculations section

- Revenue—Compute the total revenue for each set. Revenue is a function of the number of sets made and the set's selling price. The number of sets made is shown in the Changing Cells section. The selling price of a set is taken from the Constants section.
- Production Costs—Compute the total production cost for each set. This amount is a function of the number of sets made and the set's production cost. The production cost of a set is taken from the Constants section.
- Time Used (Minutes), Investment Casting—Compute the total machine time used in the investment casting stage for men's and women's sets. This number is a function of the number of sets made and the amount of investment casting time needed for a set.
- Time Used (Minutes), Tube Drawing—Compute the total machine time used in the tube drawing stage for men's and women's sets. This number is a function of the number of sets made and the amount of tube drawing time needed for a set.
- Time Used (Minutes), Assembly—Compute the total machine time used in the assembly stage for men's and women's sets. This number is a function of the number of sets made and the amount of assembly time needed for a set.

Income Statement Section

The statement shown in Figure 9-6 is the projected net income for the year. An explanation of the line items follows the figure.

	A	B	C
52	INCOME STATEMENT		
53	TOTAL REVENUE		
54	EXPENSES:		---
55	PRODUCTION COSTS		
56	SELLING, GENERAL & ADMINISTRATIVE		
57	TOTAL EXPENSES		
58	INCOME BEFORE TAXES		
59	INCOME TAX EXPENSE		
60	NET INCOME AFTER TAXES		

FIGURE 9-6 Income Statement section

- Total Revenue—This amount is computed by totaling the previously calculated revenues.
- Production Costs—This amount is computed by totaling the previously calculated production costs.
- Selling, General & Administrative—This amount can be echoed from the Constants section.
- Total Expenses—The sum of the production costs and the SG&A expenses.
- Income before Taxes—Total revenue minus total expenses.
- Income Tax Expense—This expense is zero if income before taxes is zero or less; otherwise, apply the tax rate to income before taxes.
- Net Income after Taxes—This amount is income before taxes minus income taxes.

Constraints and Running the Solver

In this part of the assignment, you determine the decision constraints. Enter the base case decision constraints using the Solver. Run the Solver and ask for the Answer Report when the Solver has found a solution that satisfies the constraints.

When you finish, print the entire workbook, including the Solver Answer Report sheet. Save the workbook by opening the File menu and selecting Save; GolfSets.xlsx is suggested as the filename. Then, to prepare for the extension case, open the File menu and select Save As to create a new spreadsheet. (**GolfSets2.xlsx** would be a good filename.)

Assignment 1B: Creating the Spreadsheet—Extension Case

What would happen to net income after taxes if you added a putter to your product line?

You calculate that you could assemble a knock-off putter in 7 minutes. The putter would look the same as the vaunted "Earthquake" putter and have the same weight and balance. (Of course, you would need a different name for your putter—you are thinking of calling it "the Kraken.")

You would sell the putter for $200. Direct production costs would be $50 per putter; these costs are primarily for the heads and shafts provided by the other manufacturer.

You think that some unused assembly time might be available to make putters. You would need to make at least 100 putters, but you do not want to take too much assembly time from the other clubs, so you think that you should make no more than 500 putters. Of course, you cannot make a fractional part of a putter.

Putters would be sold separately, not as part of a set. In your changing cells, you specify the number of putters made. The changing cells would look like Figure 9-7.

	A	B	C
1	GOLF CLUB PRODUCT MIX PROBLEM		
2			EXTENSION CASE
3	CHANGING CELLS		
4	NUMBER OF SETS OF MEN'S CLUBS		1
5	NUMBER OF SETS OF WOMEN'S CLUBS		1
6	NUMBER OF PUTTERS		1

FIGURE 9-7 Extension case Changing Cells section

Again, the goal is to maximize net income. Modify the extension case spreadsheet to handle the increased product line, and modify the constraints as needed. Run the Solver and ask for the Answer Report when the Solver finds a solution that satisfies the constraints.

When you finish, print the entire workbook, including the Solver Answer Report sheet. Save the workbook, close the file, and exit Excel.

ASSIGNMENT 2: USING THE SPREADSHEET FOR DECISION SUPPORT

You have built the base case and extension case models because you want to know how many of each product to make. You will adopt the scenario that yields the highest net income after taxes, consistent with the constraints. You will now complete the case by using the Answer Reports to gather the data needed to make the product mix decision and by documenting your recommendations in a memorandum.

Assignment 2A: Using the Spreadsheet to Gather Data

You have printed the Answer Report sheets for each scenario. Each sheet reports how many of each kind of product to make in the year. Summarize key data in a table in your memorandum. The table format is shown in Figure 9-8.

	Base Case	Extension Case	Difference
Net income after taxes (in dollars)			
Number of sets of men's clubs			
Number of sets of women's clubs			
Number of putters	0		

FIGURE 9-8 Format of table to insert in memorandum

Assignment 2B: Documenting Your Recommendation in a Memorandum

Use Microsoft Word to write a brief memorandum. State the results of your analysis and recommend which strategy to adopt: the base case or the extension case. Compare the two Answer Sheets to see if making putters merely uses slack assembly time or if it requires "stealing" assembly time from golf club sets. (You can see data for this issue in the line for the assembly time constraint.) In your memo, observe the following requirements:

- Set up your memorandum as discussed in Tutorial E.
- In the first paragraph, briefly define the situation and state the purpose of your analysis.
- Then describe your results and state your recommendations. State the logic that supports your recommendations.
- Support your statements graphically by including the summary table shown in Figure 9-8. The procedure for creating a table in Word is described in Tutorial E.

ASSIGNMENT 3: GIVING AN ORAL PRESENTATION

Your instructor may request that you present your analysis and recommendations in an oral presentation. If so, assume that you should explain your analysis and recommendations to your banker in 10 minutes or less. Use visual aids or handouts that you think are appropriate. Tutorial F explains how to prepare and give an oral presentation.

DELIVERABLES

Assemble the following deliverables for your instructor:

1. A printout of your memorandum
2. Spreadsheet printouts
3. Electronic media such as USB key or CD, which should include your Word memo and your Excel spreadsheet file

Staple the printouts together with the memorandum on top. If you have more than one .xlsx file on your electronic media, write your instructor a note that identifies the name of your model's .xlsx file.

PART 4

DECISION SUPPORT CASE USING BASIC EXCEL FUNCTIONALITY

CASE **10**

THE HOMETOWN NEWSPAPER CASH FLOW CRISIS

Decision Support Using Excel

PREVIEW

You own your city's newspaper, which is losing money and heavily in debt. Can cash flow be improved enough to avoid bankruptcy? In this case, you will use Excel to answer that question.

PREPARATION

- Review spreadsheet concepts discussed in class and in your textbook.
- Complete any exercises that your instructor assigns.
- Complete any parts of Tutorials C and D that your instructor assigns, or refer to them as necessary.
- Review file-saving procedures for Windows programs, as discussed in Tutorial C.
- Refer to Tutorial E as necessary.

BACKGROUND

Your city's newspaper has been owned by your family for many generations, and you are now the CEO. You publish a "hard copy," or printed edition, of the paper every day. The paper sells for 75 cents Monday through Saturday and for $2 on Sunday. Each day's edition is also available for free on the company's Web site. For many years, hard-copy sales have declined as Web site visits have increased. The newspaper has been losing money for many years.

You have been cutting costs and borrowing from your banker for years, trying to keep the newspaper afloat. Bank debt is now $400 million, and your bankers say they will lend no more money. Unless you can develop a plan, your paper will be forced into bankruptcy.

Your situation is typical. Newspapers in mid-sized and large U.S. cities have been failing for more than a decade. Sales of hard-copy newspapers have plummeted, and sales will continue to decline in coming years.

One reason for falling sales is that people now get the news from a variety of Web-based sources. As a result, newspapers have been forced to put their editions on the Web, free of charge. Why should someone buy a paper when it is available for free on the Web?

People also shop for items like automobiles, apartments, and real estate on the Web. Before the Internet, people would use a newspaper's Classifieds section to see what was available. Again, why pay to read ads in the paper when the information is available for free on the Web? Your newspaper now gets very little revenue from hard-copy classified advertisements.

This trend has not been a complete disaster for newspapers because they can sell advertising space in the Web editions. Thus, as people read the online version, they also see advertisements.

You expect hard-copy sales to continue to decline in the next five years, but you expect online readership to increase; this means that Web site "page views" are expected to increase. Online advertising charges are keyed to page views, and so you expect online advertising revenue to increase in the next five years.

The newspaper has 1,000 professional employees, including the writers, reporters, editors, and management staff. The professionals are not unionized. In the next five years, many professionals will retire. As a way to cut costs, you do not plan to replace retirees. In addition, you have told the professional staff that they will receive no salary or benefits increases for the foreseeable future.

The newspaper has 3,000 unionized employees, including printing-press operators, delivery staff, and janitors. In the next five years, many of these employees will retire as well. As a way to cut costs, you do not plan to replace retirees. You have bargained hard with the union, whose leaders understand the newspaper's precarious financial position. Under the current agreement, union employees will receive no hourly wage or benefits increases for the foreseeable future.

For years, your bankers have been loaning money to the newspaper to cover working capital shortages. The interest rate on your bank debt is now 15%. The collateral is the newspaper's real estate and equipment. The bankers say they will lend no more money. If the newspaper fails to make its loan payments, the banks can force you into bankruptcy, seize the real estate and equipment, and sell it for what they can get to liquidate the loans.

How can the newspaper be saved? You have a five-year plan that includes further employee cuts, a lower bank debt interest rate, and a larger family investment. Here are the main features of your plan:

- You can lay off professional employees. You think you could lay off up to 25 professionals in each of the next five years. You would be sacrificing some quality (you can get by without a book review editor, for example) but still survive. These cuts would be in addition to the planned retiree attrition.
- Your union contract allows you to lay off "noncritical" unionized employees. You think you could lay off up to 20 unionized workers in each of the next five years without causing significant labor unrest. These cuts would be in addition to the planned retiree attrition.
- The interest rate on your bank debt is currently 15%. You think you could talk your bankers into taking up to a 10% rate cut; in other words, you think they would accept a 1.5 percentage point reduction, down to 13.5%. The bankers will not like this proposal, but your promise of future business in return for the help now will keep them from foreclosing.
- You and your family will invest your own funds to cover any cash shortfall in a year, in return for added common stock. You will have to do this because the bankers have refused further lending.

The family is willing to cover a cash shortfall in a year only if the newspaper will have a cash surplus in a later year. The family is not willing to "throw good money after bad." If it looks like the five-year plan will not solve the problem, the family would rather throw in the towel now. For example, the family could accept the following five-year pattern of shortfalls and surpluses (in multiples of millions): −$50, −$25, −$25, +$50, +$60. The plan would result in a five-year cumulative total of a positive $10 million. However, the family would not accept the following five-year pattern (in millions): −$50, −$25, −$25, −$50, +$60. This plan would not work because the five-year cumulative total would be a negative $90 million.

You have been called in to create a spreadsheet that will show whether your plan can work. You need a five-year financial projection that indicates whether a combination of employee cuts, interest rate reductions, and family cash investment will stabilize the newspaper's finances. If the projection shows that finances can be improved, you will keep operating. Otherwise, you will declare bankruptcy.

ASSIGNMENT 1: CREATING A SPREADSHEET FOR DECISION SUPPORT

In this assignment, you will produce a spreadsheet that models the business problem. Then, in Assignment 2, you will write a memorandum that explains your findings and recommends a course of action. In Assignment 3, you might also be asked to prepare an oral presentation of your analysis and recommendations.

A spreadsheet has been started. *The spreadsheet skeleton is available for you to use; you can choose to type in it or not.* To access the spreadsheet skeleton, go to your data files, select Case 10, and then select **Newspaper.xlsx**. Your worksheet should include the following sections:

- Constants
- Inputs
- Summary of Key Results
- Calculations
- Cash Flow

A discussion of each section follows.

Constants Section

Your spreadsheet should include the constants shown in Figure 10-1. These constants will be used in the spreadsheet formulas. An explanation of the line items follows the figure.

	A	B	C	D	E	F	G
1	HOMETOWN NEWSPAPER CASH FLOW CRISIS						
2							
3	CONSTANTS	2010	2011	2012	2013	2014	2015
4	Hard copy decrease factor	NA	5.0%	6.0%	7.0%	8.0%	9.0%
5	Page view increase factor	NA	5.0%	7.0%	9.0%	11.0%	13.0%
6	Ad revenue per page view	NA	$0.47	$0.49	$0.51	$0.53	$0.55
7	Professional employees	NA	1,000	900	850	800	800
8	Union employees	NA	3,000	2,900	2,800	2,700	2,600
9	Daily revenue per issue	NA	$0.75	$0.75	$0.75	$0.75	$0.75
10	Sunday revenue per issue	NA	$2.00	$2.00	$2.00	$2.00	$2.00
11	Debt outstanding in year	NA	$390,000,000	$380,000,000	$370,000,000	$360,000,000	$350,000,000
12	Contract Interest rate on debt	NA	15.0%	15.0%	15.0%	15.0%	15.0%
13	Avg salary + benefits -- Professional	NA	$110,000	$110,000	$110,000	$110,000	$110,000
14	Avg salary + benefits -- Union	NA	$55,000	$55,000	$55,000	$55,000	$55,000
15	Income tax rate	NA	40.0%	40.0%	40.0%	40.0%	40.0%
16	Other revenue	NA	$1,000,000	$1,000,000	$1,000,000	$1,000,000	$1,000,000
17	General and administrative	NA	$25,000,000	$25,000,000	$25,000,000	$25,000,000	$25,000,000

FIGURE 10-1 Constants section

- Hard copy decrease factor—Each year the newspaper sells fewer hard copies. The expected rate of decrease in each year is shown. In 2011, the company expects to sell 5% fewer papers than in 2010. In 2012, the company expects to sell 6% fewer papers than in 2011, and so on.
- Page view increase factor—Web site usage, on the other hand, is expected to increase each year. The expected rate of increase in each year is shown. In 2011, the company expects 5% more page views than in 2010. In 2012, the company expects 7% more views than in 2011, and so on.
- Ad revenue per page view—Web advertisers' payments to the newspaper are based on Web page views. Advertisers pay more as views increase. The expected payments per view are shown for the next five years. Payments are expected to increase, as shown.
- Professional employees—The number of professional employees each year is expected to decrease with attrition, as shown.
- Union employees—The number of unionized employees each year is expected to decrease with attrition, as shown.
- Daily revenue per issue—The hard copy of the newspaper sells for 75 cents Monday through Saturday. The newspaper does not plan to increase this price in the next five years.
- Sunday revenue per issue—The hard copy of the Sunday newspaper sells for $2. The newspaper does not plan to increase this price in the next five years.
- Debt outstanding in year—The banks require that $10 million be repaid each year. In 2011, $390 million will be owed. In 2012, $380 million will be owed, and so on.
- Contract interest rate on debt—The rate on bank debt is 15% each year, as shown.
- Avg salary + benefits – professional—In the old days, management was very generous with the staff. The average salary for professionals is $110,000 a year, which is high in the industry. This amount is not expected to increase in the next five years.
- Avg salary + benefits – union—The average salary for union employees is $55,000 a year. This amount is not expected to increase in the next five years.
- Income tax rate—The tax rate on income before taxes is expected to be 40% in each year, as shown.

- Other revenue—Small amounts of revenue are earned as interest on cash in the bank, classified ads in the hard copy of the newspaper, and so on. The amount is not expected to change in the next five years.
- General and administrative—General business expenses are estimated to be $25 million per year.

Inputs Section

Your worksheet should include the three inputs shown in Figure 10-2; these inputs will be used by the spreadsheet's formulas. Your instructor may tell you to apply conditional formatting to the input cells so that out-of-bounds values are highlighted in some way. An explanation of the inputs follows the figure.

	A	B	C	D	E	F	G
19	INPUTS	2010	2011	2012	2013	2014	2015
20	Number of salaried layoffs in year (0-25)	NA					
21	Number of union layoffs in year (0-20)	NA					
22	Interest rate reduction in year (0-10%)	NA					

FIGURE 10-2 Inputs section

- Number of salaried layoffs in year (0–25)—You think you can lay off up to 25 professionals a year, in addition to the head-count reductions shown in the Constants section. Enter a reduction for each of the five years. If you want to see the effect of laying off 10 professionals a year, enter the sequence 10, 10, 10, 10, 10. If you want to lay off 10 employees in the first three years, but none in the last two years, enter the sequence 10, 10, 10, 0, 0.
- Number of union layoffs in year (0–20)—You think you can lay off up to 20 unionized employees a year, in addition to the head-count reductions shown in the Constants section. Enter a reduction for each of the five years.
- Interest rate reduction in year (0–10%)—You think the banks will take a lower rate of interest in each year. You think the reduction could be between 0 and 10% of the expected rate in the year. (The expected rate is shown in the Constants section.) A 10% reduction would imply a 1.5 percentage point reduction (10% of the 15% rate), resulting in an interest rate of 13.5% for the year. If you enter 5%, the interest rate reduction for the year would be 0.75 of a percentage point, resulting in a rate of 14.25% for the year.

Summary of Key Results Section

Your worksheet should include the key results shown in Figure 10-3. The values are echoed from another section of your spreadsheet. An explanation of the results follows the figure.

	A	B	C	D	E	F	G
24	SUMMARY OF KEY RESULTS	2010	2011	2012	2013	2014	2015
25	Cumulative Cash Flow	NA					

FIGURE 10-3 Summary of Key Results section

- Cumulative Cash Flow—Your plan can result in a deficit in some years and surpluses in later years. In any particular year, what is the "cumulative" position? For example, if the paper has a $10 million deficit in year 1 and a $5 million surplus in year 2, the cumulative deficit will be $5 million after two years. If you have a $10 million surplus in year 3, the cumulative surplus at that point will be $5 million. This value is computed elsewhere in your spreadsheet and should be echoed here.

Calculations Section

This section calculates the values shown in Figure 10-4. Values are calculated by formula, not hard-coded. These values are used in other calculations and in the cash flow section that follows. An explanation of the line items follows the figure.

	A	B	C	D	E	F	G
27	**CALCULATIONS**	**2010**	**2011**	**2012**	**2013**	**2014**	**2015**
28	Mon-Sat Hard copy sales	145,000,000					
29	Sunday hard copy sales	28,000,000					
30	Mon-Sat Hard copy revenue	NA					
31	Sunday hard copy revenue	NA					
32	Page views per year	320,000,000					
33	Page view revenue	NA					
34	Adjusted Interest rate in year	NA					
35	Professional employees, post layoffs	NA					
36	Union employees, post layoffs	NA					
37	Professional salary + benefits	NA					
38	Union salary + benefits	NA					

FIGURE 10-4 Calculations section

- Mon-Sat Hard copy sales—Hard-copy sales are a function of sales in the prior year and the hard-copy decrease factor for the year. The factor is taken from the Constants section.
- Sunday hard copy sales—Hard-copy sales on Sunday are a function of Sunday sales in the prior year and the hard-copy decrease factor for the year. The factor is taken from the Constants section.
- Mon-Sat Hard copy revenue—This amount is a function of Monday-Saturday hard-copy sales and the daily revenue per issue. The daily revenue per issue is taken from the Constants section.
- Sunday hard copy revenue—This amount is a function of Sunday hard-copy sales and the Sunday revenue per issue. The revenue per issue is taken from the Constants section.
- Page views per year—This value is a function of the prior year's Web page views and the page view increase factor for the year. The latter value is taken from the Constants section.
- Page view revenue—This amount is a function of the page views in the year and the revenue per page view in the year. The latter value is taken from the Constants section.
- Adjusted interest rate in year—This rate is a function of the contract interest rate on debt and the interest rate reduction. The latter value is taken from the Inputs section.
- Professional employees, post layoffs—This amount is a function of the number of professional employees after attrition in the year (from the Constants section) and the salaried employee layoffs in the year (from the Inputs section).
- Union employees, post layoffs—This amount is a function of the number of union employees after attrition in the year (from the Constants section) and the union employee layoffs in the year (from the Inputs section).
- Professional salary + benefits—This amount is a function of the number of professionals employed after layoffs (from the Calculations section) and the average professional salary and benefits paid in the year (from the Constants section).
- Union salary + benefits—This amount is a function of the number of unionized employees after layoffs (from the Calculations section) and the average union salary and benefits paid in the year (from the Constants section).

Cash Flow Section

This section is the spreadsheet body, as shown in Figure 10-5. Cash inflows and cash outflows are summarized into a net cash flow for the year (a deficit or surplus) and a cumulative cash flow to date. You hope that you can find a way to generate a positive cumulative cash flow at the end of 2015. An explanation of the line items follows the figure.

	A	B	C	D	E	F	G
40	**CASH FLOW**	2010	2011	2012	2013	2014	2015
41	Hard Copy Revenue	NA					
42	Page View Revenue	NA					
43	Other Revenue	NA					
44	Total Revenue	NA					
45	Salary and benefits	NA					
46	General and Administrative	NA					
47	Total Costs	NA					
48	Income before interest and taxes	NA					
49	Interest expense	NA					
50	Income before taxes	NA					
51	Income tax expense	NA					
52	Income after taxes	NA					
53	Less: debt repayment in year	NA	$10,000,000	$10,000,000	$10,000,000	$10,000,000	$10,000,000
54	Cash flow in year	NA					
55	Cumulative Cash Flow	NA					

FIGURE 10-5 Cash Flow section

- Hard Copy Revenue—The total of weekday and Sunday hard-copy revenues, both of which are taken from the Calculations section.
- Page View Revenue—This amount is taken from the Calculations section and can be echoed here.
- Other Revenue—This amount is taken from the Constants section and can be echoed here.
- Total Revenue—The total of hard copy revenue, page view revenue, and other revenue for the year.
- Salary and benefits—The total amount paid both for professional and union compensation, which are taken from the Calculations section.
- General and Administrative—This amount is taken from the Constants section and can be echoed here.
- Total Costs—The sum of salary and benefits and general and administrative costs.
- Income before interest and taxes—The difference between total revenue and total costs.
- Interest expense—A function of the debt outstanding in the year (from the Constants section) and the adjusted interest rate in the year (from the Calculations section).
- Income before taxes—The difference between income before interest and taxes and interest expense.
- Income tax expense—Taxes are zero if income before taxes is zero or negative. Otherwise, apply the income tax rate (from the Constants section) to income before taxes to compute income tax expense.
- Income after taxes—The difference between income before taxes and income tax expense.
- Less: debt repayment in year—This amount is a $10 million payment to the banks each year, as shown.
- Cash flow in year—The difference between income after taxes and the year's debt repayment.
- Cumulative Cash Flow—The net effect of the cash flows in the years to date.

ASSIGNMENT 2: USING THE SPREADSHEET FOR DECISION SUPPORT

You will now complete the case by using the spreadsheet to gather data needed to determine whether a positive cumulative cash flow can be generated using your plan. You will also document your findings and recommendations in a memorandum.

Assignment 2A: Using the Spreadsheet to Gather Data

You have built the spreadsheet to create "what-if" scenarios using the plan's input values. You want to know if a combination of inputs will generate a positive cumulative cash flow in five years. If you can find the proper combination, the plan will be implemented in an effort to save the newspaper. Otherwise, the newspaper must declare bankruptcy.

Here are some guidelines for your "what-if" analysis, in the form of questions that your spreadsheet can answer:

1. Is it possible to generate a positive cumulative cash flow after 2015 by laying off only professionals? In other words, you would not lay off any union workers or adjust the interest rate in this scenario.
2. Is it possible to generate a positive cumulative cash flow after 2015 by laying off only union workers? In other words, you would not lay off any professional workers or adjust the interest rate in this scenario.
3. Is it possible to generate a positive cumulative cash flow after 2015 by only adjusting the interest rate? You would not lay off professional or union employees in this scenario.
4. Can a positive cumulative cash flow be generated after 2015 only by using the maximum input values of 25 professionals laid off, 20 union workers laid off, and 10% rate reduction?
5. What combination of input values produces the maximum cumulative cash flow in 2015?
6. Is there no combination of inputs that generates a positive cumulative cash flow in 2015? In other words, is it necessary to declare bankruptcy?

Run your what-if analysis. You could use the Scenario Manager to organize the effort, but it is not required. You could also enter values manually by writing the results on paper. When you finish gathering the data, print the worksheet using any set of inputs. Then save the spreadsheet by opening the File menu and selecting Save.

Assignment 2B: Documenting your Findings and Recommendations in a Memorandum

Use Microsoft Word to document your findings in a memorandum that answers the questions in the previous section. In your memo, observe the following requirements:

- Your memorandum should have proper headings such as Date, To, From, and Subject. You can address the memorandum to your family members and your bankers. Set up the memorandum as discussed in Tutorial E.
- Briefly outline the situation. However, you need not provide much background—you can assume that readers are familiar with the problem.
- Answer the key questions in the body of the memorandum.
- Support your claims by showing key results in a table. Tutorial E describes how to create a table in Microsoft Word.

ASSIGNMENT 3: GIVING AN ORAL PRESENTATION

Your instructor may request that you present your analysis and results in an oral presentation. If so, assume that the family and your bankers are impressed by your analysis. Prepare to talk to the group for 10 minutes or less. Use visual aids or handouts that you think are appropriate. Tutorial F explains how to prepare and give an oral presentation.

DELIVERABLES

Assemble the following deliverables for your instructor:

1. Printout of your memorandum
2. Spreadsheet printouts
3. Electronic media such as USB key or CD that should include your Word file and Excel file

Staple the printouts together with the memorandum on top. If you have more than one .xlsx file on your electronic media, write your instructor a note that identifies your spreadsheet model's .xlsx file.

PART 5

INTEGRATION CASES USING ACCESS AND EXCEL

CASE **11**

THE CREDIT DEFAULT SWAP DECISION

Decision Support with Access and Excel

PREVIEW

A large pension fund invested in triple-A rated bonds and used your firm to insure the payment of the bonds' principal and interest. The bonds' collateral is a group of residential mortgages, and now the pension fund might want to collect on the insurance. In this case, you will use Access and Excel to analyze the quality of the collateral and determine whether your firm must pay the insurance claim.

PREPARATION

- Review database and spreadsheet concepts discussed in class and in your textbook.
- Complete any exercises that your instructor assigns.
- Complete any parts of Tutorials B, C, and D that your instructor assigns, or refer to them as necessary.
- Review file-saving procedures for Windows programs, as discussed in Tutorial C.
- Refer to Tutorial E as necessary.

BACKGROUND

Your investment banking firm is always looking for profitable investments. In recent years, your firm has issued credit default swaps, which are a form of investment insurance, to increase revenue and improve the bottom line.

Swaps, however, now look risky to your firm. To understand why, you need to understand the situation and the following groups:

- Home owners—People generally need to take out a mortgage when they buy a house. A mortgage is a loan that must be repaid over a period of time (usually 15 or 30 years). For example, a person might buy a house for $200,000, make a $20,000 down payment, and take out a 15-year mortgage at their local bank to pay for the balance. The person now owns the house but owes $180,000 to the bank. Every month for 15 years, the person makes a payment to the bank. The payment is partly interest and partly loan principal. At the end of 15 years, the loan is paid off.
- Local bank—To a bank, a loan is a productive asset that earns interest income as payments are made over 15 or 30 years. However, sometimes a bank would rather realize the full value of the loan immediately rather than over time. For instance, banks must "service" their loans by keeping track of payments made and how much customers owe, and some banks would rather not be bothered with that burden. Also, a bank might have an immediate use for the full cash value of a loan. In that case, the mortgage's promissory note—which is a negotiable instrument—can be sold for any amount the market will bear. To use the previous example, the $180,000 mortgage could be sold to an investor who wants a 15-year stream of payments.
- Structured investment vehicle (SIV)—An SIV is a company that buys mortgages from local banks. The mortgages become the SIV's assets, so the monthly payments flow to the SIV.

The SIV gets the cash to buy mortgages by issuing a bond called a collateralized debt obligation (CDO), which is bought by a wealthy investor looking for an investment that pays interest. For example, an SIV might buy 200 fifteen-year mortgages from banks in the region. The total value of all the loans could be $50,000,000 and the average interest rate could be 6%. Simultaneously, the SIV issues a 15-year bond to a wealthy investor that pays 5%. Every month, the 200 home owners send their payments to the SIV, and the SIV sends a payment to the wealthy investor to cover the bond's monthly principal and interest. The difference in rates (here, 6% versus 5%) is what allows the SIV to prosper. Note that if home owners fail to make their monthly payments, the SIV might have trouble making its payments to the wealthy investor. SIV payments might be late as it waits for sufficient cash to flow in.

- Investors—The wealthy investor that buys the SIV's bond could be a pension fund, a mutual fund, an investment bank, some other finance company, or even an individual. The 200 home mortgages from the previous example are the bond's collateral. If the SIV defaults (fails to make payments), the investor can foreclose on the homes. The bond is called a collateralized debt obligation because the obligation is backed up by collateral. Investors do not want to foreclose on houses, however, so they seek two other forms of protection: the blessing of a credit rating agency and insurance in the form of a credit default swap.
- Credit rating agency—The SIV will have a credit rating agency review its mortgage deals. If the rating agency sees problems, it will issue a poor rating. If the agency sees no problems, it issues a good rating, which gives the investor confidence that SIV payments will be made.
- Insurer—The investor is willing to pay a fee for insurance against the possibility that the SIV will default on the bond's payments. This fee is revenue to the insurer. If the SIV's payments are made, the insurer has no expense; but if a default occurs, the insurer must send a check to the investor for some or all of the bond's full value. The insurance instrument is called a credit default swap. The insuring company can be an actual insurance company, an investment bank, or another kind of financial firm.

In fact, this kind of complex transaction has been common in recent years. You work for an investment bank that has earned tremendous fees for credit default swaps in the past few years. Your firm has an ongoing relationship with an SIV called Desert Storm Investments and a pension fund named Midwest Wealth Management.

Three years ago, Desert Storm worked with a number of Western mortgage banks to put together a CDO (code-named "Curly"). Pacific Surf Analytics, a well-known credit agency, gave its highest rating (AAA) to Curly's collateral, which consisted of 200 residential home mortgages. The bond was sold to Midwest Wealth Management, a large pension fund that paid your company $2 million for the credit default swap. Desert Storm has faithfully made monthly payments on Curly to Midwest.

Two years ago, Desert Storm again worked with its mortgage banks to put together another CDO (code-named "Larry"). Pacific Surf Analytics gave its highest rating to Larry's collateral, which also consisted of residential home mortgages. The bond was sold to Midwest Wealth Management, which paid your company $2 million for another credit default swap. Desert Storm has been late making monthly payments on Larry to Midwest.

One year ago, Desert Storm again worked with its mortgage banks to create another CDO (code-named "Moe"). Pacific Surf Analytics gave its highest rating to Moe's collateral, which was another batch of residential home mortgages. The bond was sold to Midwest Wealth Management, which paid your company $2 million for another credit default swap. Desert Storm has been late making monthly payments on Moe to Midwest.

You have recently been promoted to an analyst's position in the credit default unit at your firm. You have no experience with this kind of investment product and the related analysis, so you investigate how your predecessor handled credit default swaps. You learn two things that make you uneasy:

1. She did not actually investigate the details of the CDO's underlying mortgages, but assumed that the credit agency Pacific Surf Analytics did the investigation. In fact, your predecessor did not even have a list of the home mortgages that backed up the insured CDOs.
2. Your contact at Pacific Surf Analytics says that her company does not look at each mortgage in the CDO. She said, "A mortgage contract has a lot of pages. There are hundreds of mortgages in a CDO. How can I read all that? Do you think I have time to investigate each home owner's credit scores?"

You learn that Pacific Surf Analytics assesses credit worthiness using a computer model that includes some high-level macroeconomic data for the region. Your contact says that the model generally predicts good results for mortgage-backed investments in your region.

Your contact at Pacific Surf Analytics explains some of the basics of real estate mortgage credit analysis. Assume, she says, that a man puts $20,000 down on a $200,000 house, taking a 7% mortgage for $180,000. Monthly payments are $1,800. After a few years, the man has some economic problems and has trouble making the monthly payment. However, there are three reasons to be optimistic. (1) In most years, U.S. real estate values have gone up, sometimes dramatically. Assume that the market value of the man's house is now $250,000, an increase of $50,000 from the purchase price. The man can borrow some percentage of that increase using a "home equity" loan; the collateral is the value of the house, which now covers both the first and second mortgages. (2) Perhaps interest rates have fallen, say to 5%. The man can now refinance his $180,000 loan so that his monthly payment is lower and more affordable. (3) Banks do not foreclose after a couple of missed payments. Therefore, the man may be in arrears, but as times improve he probably will make payments again.

In summary, your contact says that the credit rating agency is very comfortable with CDOs that have real estate mortgage collateral.

However, you know that mortgage rates have not changed much in the last few years, and refinancings are not as common as they used to be. Also, you know that real estate values have not been skyrocketing lately.

Furthermore, you've read your company's credit default insurance contract. The investor can claim the insurance when the value of the collateral has been "significantly impaired." In other words, the investor does not have to wait until the SIV's monthly payments dry up and 200 homes are foreclosed. The contract says that the investor can file an insurance claim if SIV payments are late. The claim would have to be paid if more than 75% of home owners are two or more months in arrears with their payments to the SIV. Such conditions would demonstrate significant impairment according to the contract.

Of course, the investor does not know which home owners are late and which are not. However, SIV payments to the investor have been late on the Larry and Moe CDOs. Are home owners not making their mortgage payments to the SIV? You are worried that the investor will lose patience and file claims on the two CDOs under the terms of your credit default swaps. Each CDO is a $35 million bond!

You ask Desert Storm to give you paperwork on each loan that supports the Curly, Larry, and Moe CDOs. Your assistant laboriously enters the data into an Access database file called **Loans.accdb**. The database includes three tables; the table Year1Data contains the records for the Curly CDO, which was formed in the first year of Desert Storm's relationship with the investor Midwest Wealth. Figure 11-1 shows the first few of the table's 200 records.

Year1Data

Loan Number	Loan Value	Appraised	Current Value	No Doc	FICO	In Arrears
1001	$82,500	$110,000	$84,175	NO	630	NO
1002	$315,400	$380,000	$348,712	NO	730	NO
1003	$91,200	$120,000	$102,648	NO	750	NO
1004	$193,600	$220,000	$199,290	NO	560	YES
1005	$251,100	$270,000	$242,151	NO	580	YES
1006	$272,600	$290,000	$266,175	NO	590	NO
1007	$281,600	$320,000	$291,928	NO	790	NO
1008	$135,000	$180,000	$159,432	NO	740	NO

FIGURE 11-1 Data for Curly CDO's loans

Field definitions are as follows:

- Loan Number—You assign a unique number to each loan. This is the table's primary key field.
- Loan Value—The amount borrowed to buy the house.
- Appraised—The selling price of the house when it was purchased. This price is assumed to be the appraised value at the time of purchase. For example, the house in loan 1001 was sold for $110,000. The owner's down payment was $27,500, and the new owner borrowed $82,500.

- Current Value—The current market value of the house *at the time you are doing the analysis*. Notice that the value of the house in loan 1001 has declined in the three years under review.
- No Doc—Lenders usually require an applicant to document the loan application thoroughly. For example, prospective borrowers must show that they have a job and a certain income. Lenders will require borrowers to submit documents such as tax returns and pay stubs to prove their claims. However, in recent years some lenders have just taken the applicant's word for his claims. In other words, they have made loans without requiring documentation; such loans are called "No Doc" loans. Loan 1001 was *not* a No Doc loan—the bank did require the applicant to prove claims about a job and income. (If the loan had been a No Doc loan, the field's value would have been YES.)
- FICO—The loan applicant's FICO score at the time of the loan. A FICO score is a measure of personal credit worthiness. FICO scores range from 300 to 850. The higher the score is, the better.
- In Arrears—This value shows whether the owner is up to date in monthly mortgage payments. A value of NO means the borrower is up to date. A value of YES means that the borrower is *two or more* mortgage payments behind. Note that being in arrears does not mean that the house is in foreclosure.

The table Year2Data contains the records for the Larry CDO, which was formed in the second year of Desert Storm's relationship with Midwest Wealth. Figure 11-2 shows the first few of the table's 200 records.

Year2Data

Loan Number	Loan Value	Appraised	Current Value	No Doc	FICO	In Arrears
2001	$77,550	$86,000	$69,000	YES	640	NO
2002	$302,784	$318,000	$289,000	YES	480	YES
2003	$91,200	$97,000	$81,000	YES	500	YES
2004	$185,856	$197,000	$179,000	YES	630	NO
2005	$238,545	$262,000	$238,000	NO	630	NO
2006	$267,148	$290,000	$254,000	YES	400	YES
2007	$281,600	$331,000	$243,000	YES	550	YES
2008	$133,650	$146,000	$116,000	YES	410	YES

FIGURE 11-2 Data for Larry CDO's loans

The field definitions are the same as those in Figure 11-1.

The table Year3Data contains the records for the Moe CDO, which was formed in the third year of Desert Storm's relationship with Midwest Wealth. Figure 11-3 shows the first few of the table's 200 records.

Year3Data

Loan Number	Loan Value	Appraised	Current Value	No Doc	FICO	In Arrears
3001	$69,550	$75,550	$66,740	NO	620	NO
3002	$302,784	$310,784	$273,540	YES	470	YES
3003	$85,200	$92,200	$78,020	YES	460	YES
3004	$185,856	$189,856	$170,140	YES	610	NO
3005	$228,545	$236,545	$225,600	YES	600	NO
3006	$260,148	$270,148	$240,640	YES	380	YES
3007	$280,600	$283,600	$230,300	YES	520	NO
3008	$123,650	$131,650	$110,920	YES	380	YES

FIGURE 11-3 Data for Moe CDO's loans

The field definitions are the same as those in Figure 11-1.

You need to use Access and Excel to see whether payments on the credit default swaps would be required if the investor filed a claim. *The database is available for you to use; you can choose to type in it or not*. To access the database, go to your data files, select Case 11, and then select **Loans.accdb**.

ASSIGNMENT 1: MAKING QUERIES IN ACCESS

In this assignment, you will design and run three queries.

Year 1 Query

Create a query that lists all data for the Curly CDO's loans and computes two ratios for each record: the loan value to appraised value ratio and the loan value to current market value ratio. Your output should look like that in Figure 11-4 (only the first few of the 200 records are shown). Name the query Year1Query.

Year1Query

Loan Number	Loan Value	Appraised	Current Value	No Doc	FICO	In Arrears	Loan to Appraised	Loan to Current
1001	$82,500	$110,000	$84,175	NO	630	NO	0.750	0.980
1002	$315,400	$380,000	$348,712	NO	730	NO	0.830	0.904
1003	$91,200	$120,000	$102,648	NO	750	NO	0.760	0.888
1004	$193,600	$220,000	$199,290	NO	560	YES	0.880	0.971
1005	$251,100	$270,000	$242,151	NO	580	YES	0.930	1.037
1006	$272,600	$290,000	$266,175	NO	590	NO	0.940	1.024
1007	$281,600	$320,000	$291,928	NO	790	NO	0.880	0.965
1008	$135,000	$180,000	$159,432	NO	740	NO	0.750	0.847

FIGURE 11-4 Year 1 query

The Loan to Appraised ratio shows the deposit that was made on the house. For example, if the ratio is less than 0.90, a deposit was made for more than 10% of the value. To format the ratio's fields, right-click the column, select Properties, and then select Standard format and 3 decimals.

Year 2 and Year 3 Queries

Create a query that lists all data for the Larry CDO's loans and computes the same two ratios for the Larry CDO's loans as the ratios for the Year 1 query. Name the query Year2Query. Next, create a query that lists all data for the Moe CDO's loans and computes the same two ratios for the loans. Name the query Year3Query. The format of the output is the same as in the Year 1 query.

When you finish the queries, save and close the **Loans.accdb** file.

ASSIGNMENT 2: USING EXCEL FOR DECISION SUPPORT

In this assignment, you will import your three Access queries into Excel worksheets and then develop information about the possible insurance claim.

Importing Queries

Open a new file in Excel and save it as **Loans.xlsx**. Then import the Year1Query data into Excel. Click the Data tab, select Get External Data, and then select From Access. Specify the Access filename, the query name, and where to place the data in Excel (cell A1 is recommended). Rename the worksheet Year1.

The data will come into Excel as an Excel data table. You should change the table to a regular data range at this point by clicking a cell in the table, clicking the Design tab, and selecting Convert to Range in the Tools group. Year 1 data should look like Figure 11-5.

	A	B	C	D	E	F	G	H	I
1	Loan Number	Loan Value	Appraised	Current Value	No Doc	FICO	In Arrears	Loan to Appraised	Loan to Current
2	1001	82500	110000	84175	NO	630	NO	0.750	0.980
3	1002	315400	380000	348712	NO	730	NO	0.830	0.904
4	1003	91200	120000	102648	NO	750	NO	0.760	0.888
5	1004	193600	220000	199290	NO	560	YES	0.880	0.971
6	1005	251100	270000	242151	NO	580	YES	0.930	1.037
7	1006	272600	290000	266175	NO	590	NO	0.940	1.024
8	1007	281600	320000	291928	NO	790	NO	0.880	0.965
9	1008	135000	180000	159432	NO	740	NO	0.750	0.847

FIGURE 11-5 Year 1 data imported

Note that you may need to format the two loan ratio columns to three decimal numbers.

Repeat the preceding instructions to import the Year2Query and Year3Query data. Use separate sheets and name them Year2 and Year3, respectively.

Further Calculations

You want to know if a loan is a so-called "subprime" loan and if a loan is "under water." A loan is subprime if the borrower's FICO score was less than 620 *and* the loan to appraised ratio is greater than 90%. In other words, the down payment was less than or equal to 10%, meaning that the borrower does not have good credit and did not make a good down payment. An under-water loan is one in which the initial loan value is greater than the current market value. An under-water loan has a loan to current ratio greater than 1.

In the Year1 sheet, create column headings for SubPrime and Underwater, and then enter formulas to compute these values. The top part of the Year1 worksheet would look like Figure 11-6.

	A	B	C	D	E	F	G	H	I	J	K
1	Loan Number	Loan Value	Appraised	Current Value	No Doc	FICO	In Arrears	Loan to Appraised	Loan to Current	SubPrime	Underwater
2	1001	82500	110000	84175	NO	630	NO	0.750	0.980	NO	NO
3	1002	315400	380000	348712	NO	730	NO	0.830	0.904	NO	NO
4	1003	91200	120000	102648	NO	750	NO	0.760	0.888	NO	NO
5	1004	193600	220000	199290	NO	560	YES	0.880	0.971	NO	NO
6	1005	251100	270000	242151	NO	580	YES	0.930	1.037	YES	YES
7	1006	272600	290000	266175	NO	590	NO	0.940	1.024	YES	YES
8	1007	281600	320000	291928	NO	790	NO	0.880	0.965	NO	NO

FIGURE 11-6 Year1 sheet with subprime and under-water calculations

Perform the same calculations in sheets Year2 and Year3.

You now want to convert the data ranges back to data tables. To do so, select the entire data range, including the header line. In the Home tab Styles group, select Format as Table. Pick a light style. The Design tab will be activated automatically. In the Table Style Options group, select Total Row. Repeat these instructions for each of the three worksheets.

Using Data Tables and Pivot Tables to Gather Data

You now want to use the data tables and pivot tables to gather data. (Consult Tutorial E if you need help using data tables and pivot tables.)

Data Table Analysis

Use your data tables to complete the three tables shown in Figures 11-7, 11-8, and 11-9. You will incorporate these tables into your memorandum at the end of the case.

Use the totals line in the data tables to gather the data shown in Figure 11-7.

	Year 1 Loans	Year 2 Loans	Year 3 Loans
Average loan value			
Average appraised value			
Average current value			
Average FICO score			
Average loan to appraised ratio			
Average loan to current ratio			

FIGURE 11-7 Average data

Use the totals line in the data tables to gather the data shown in Figure 11-8.

	Year 1 Loans	Year 2 Loans	Year 3 Loans
Count of No Doc loans			
Count of loans in arrears			
Count of subprime loans			
Count of under-water loans			

FIGURE 11-8 Count data

In each year, the SIV selected 200 loans to back up the CDO. After looking at the averages and the counts in the tables, you should consider the following questions:

- Did the SIV select different kinds of loans (or borrowers) in the three years? In other words, did the real estate market change so that different loans were selected?
- Does it appear that lenders changed lending standards so that less capable payers were given loans?
- Does it appear that market values have increased enough to allow home owners to borrow under home equity loans?
- For each CDO, what percentage of loans is in arrears?

Use the totals line in the data tables to gather the data shown in Figure 11-9. This data is *for loans that are in arrears*.

	Year 1 Loans	Year 2 Loans	Year 3 Loans
Count of loans in arrears			
Average loan to appraised ratio			
Average loan to current ratio			
Count of subprime loans			
Count of under-water loans			
Count of both subprime and under-water loans			

FIGURE 11-9 Data for loans in arrears

Note that the count of loans in arrears should agree with the same count in Figure 11-8. Note also that you can set a double filter in the top line of the table. In the header line, set In Arrears to YES and then set a second filter (for example, set SubPrime to YES). Then use the totals row to gather data.

While looking at the data, you should try to see what borrower characteristics seem to correlate with payment slowness. Are loans that are in arrears likely to be subprime loans? Under-water loans? Both? Does a low deposit predict slow pay?

Pivot Table Analysis

In each worksheet, create three pivot tables.

The first pivot table should show the average loan to appraised ratios for loans that are in arrears and loans that are not. You should tell Excel to put the pivot table in the same worksheet you developed earlier, with the upper-left corner at A205. Examine the data and consider the following question: Does the size of the deposit predict very well whether a borrower will go into arrears?

The second pivot table should show the average loan to current ratios for loans in arrears and loans that are not. You should tell Excel to put the pivot table in the same worksheet, with the upper-left corner at A210. While looking at the data, you should consider the following question: Does the ratio predict very well whether a borrower will go into arrears?

The third pivot table should show the average FICO score to appraisal ratios for loans in arrears and loans that are not. You should tell Excel to put the pivot table in the same worksheet, with the upper-left corner at A215. While looking at the data, you should consider the following question: Does the borrower's FICO score predict very well whether a borrower will go into arrears?

While looking at all the pivot table data, consider the following question: Which of the three measures seems to be the best predictor of whether a borrower will go into arrears?

ASSIGNMENT 3: DOCUMENTING FINDINGS IN A MEMORANDUM

In this assignment, you write a memorandum in Microsoft Word that documents your findings and recommendations. You should describe the business situation briefly. Summarize the nature of the collateral in the three years and indicate how things have changed. You should create Word tables like those in Figures 11-7, 11-8, and 11-9 as addendums to your memorandum.

You should be able to state whether your company will be liable for a claim under any of the three credit default swaps you have written with the investor.

Note that a successful claim would be for the amount of the investor's bond, which would be about the same as the total of the initial loans in the CDO's collateral. If you are close to being liable for a claim under a credit default swap and you think that the situation may worsen as time goes on (in other words, more borrowers will go into arrears), answer the following questions: Should your company try to help the home borrowers make payments to the SIV? If so, what sort of help is needed, and how much? Would a better plan be to preemptively settle with the investor for a lesser amount (say, $10 million) to get out of the insurance contract?

In your memo, observe the following requirements:

- Your memorandum should have proper headings such as Date, To, From, and Subject. You can address the memorandum to your manager. Set up the memorandum as discussed in Tutorial E.
- Briefly outline the situation. However, you need not provide much background about the business—you can assume that readers are familiar with the real estate business, CDOs, and credit default swaps.
- Answer the preceding questions in the body of the memorandum.
- Support your claims by showing important results in tables. Tutorial E describes how to create a table in Microsoft Word.

DELIVERABLES

Assemble the following deliverables for your instructor:

1. Printout of your memorandum
2. Spreadsheet printouts
3. Electronic media such as USB key or CD, which should include your Word file, Access file, and Excel file

Staple the printouts together with the memorandum on top. If you have more than one .xlsx file or .accdb file on your electronic media, write your instructor a note that identifies this assignment's files.

CASE **12**

THE MUSIC PREDICTION SYSTEM

Designing a Relational Database to Create Tables, Queries, and Decision Support Using Access and Excel

PREVIEW

In this case, you will design a relational database in Access for a Web-based music listening company. Next, you will create the tables and populate them with data. From the database, you will create three queries to list the number of songs available in the system's library by genre, to list a given customer's favorite songs, and to search for songs. Next, you will import data into Excel from the database, use the least squares method to predict which songs a given customer will like, and recommend songs to that customer.

PREPARATION

- Review spreadsheet and database concepts discussed in class and your textbook.
- Complete any part of Database Design Tutorial A that your instructor assigns.
- Complete any part of Tutorials B, C, or D that your instructor assigns, or refer to the tutorials as necessary.
- Refer to Tutorials E and F as necessary.

BACKGROUND

You have just landed a job with a new company that provides music via the Internet. The company has a large playlist of old and current music that it makes available for streaming on the Web. The company earns income by selling advertising on the site. To gain more advertising revenue and customers, the company wants to start a system in which customers can receive recommendations for music based on their past selections. Before migrating this new system to the Web, the company would like to have a prototype built to test various features. Your expertise in Access and Excel is a perfect fit for the company because it can use both Windows applications to simulate the new system.

Before you begin to design the database, you must keep a number of parameters in mind. The company has a huge library of songs, and it keeps track of song information such as title, length, artist, CD name, and music genre. Obviously you must record customer information such as name, address, phone number, e-mail address, and billing information. You will also record which songs each customer listens to and how many times they listen to each song. This information is important for predicting which future songs the customer might like.

Once your database is complete and populated with data, you can begin the analysis portion of the project. The company would like to be able to extract information from the database, and you suggest that the best method is to use queries. The first query should report how many songs from each music genre are included in the company library. The output of this query can be used for marketing purposes. Second, customers often would like to search for songs by a particular artist. You suggest that a parameter query would satisfy this request. Finally, the company would like a listing of customers' favorite songs and the number of times they play those songs; this is vital information for future song prediction.

The query information and database table should be imported into Excel, where a least squares fit of the data can be run. The least squares fit is a mathematical procedure that finds the best fitting curve to a set of points. From this fit, you can make certain predictions. Using attributes from the Access analysis, you can predict whether a customer might be interested in listening to a particular song. Note that least squares fit uses only numeric data, so your design must include numeric coding of the genres.

The owners of the company realize that your system will simply be a model or prototype of the final system. Obviously, they want to attract a huge number of customers, and your Access database and analysis of one customer will not be sufficiently robust. However, your job is important to lay the groundwork for the new company's future information systems.

ASSIGNMENT 1: CREATING THE DATABASE DESIGN

In this assignment, you will design your database tables using a word-processing program. Pay close attention to the tables' logic and structure. Do not start developing your Access code in Assignment 2 before getting feedback from your instructor on Assignment 1. Keep in mind that you will need to examine the requirements in Assignment 2 to design your fields and tables properly. It's good programming practice to look at the required outputs before designing your database. When designing the database, observe the following guidelines:

- First, determine the tables you'll need by listing the name of each table and the fields it should contain. Avoid data redundancy. Do not create a field if it can be created by a "calculated field" in a query.
- You'll need at least one transaction table. Avoid duplicating data.
- Document your tables using the Table feature of your word processor. Your word-processed tables should resemble the format shown in Figure 12-1.
- You must mark the appropriate key field(s) by entering an asterisk (*) next to the field name. Keep in mind that some tables might need a compound primary key to uniquely identify a record within a table.
- Print the database design.

Table Name	
Field Name	Data Type (text, numeric, currency, etc.)
...	...
...	...

FIGURE 12-1 Table design

NOTE

Have your design approved before beginning Assignment 2; otherwise, you may need to redo Assignment 2.

ASSIGNMENT 2: CREATING THE DATABASE, CREATING QUERIES, USING EXCEL FOR DECISION SUPPORT, AND IMPORTING DATA

In this assignment, you will first create database tables in Access and populate them with data. Next, you will create three queries to export to Excel. You'll analyze the data in Excel and return it to Access.

Assignment 2A: Creating Tables in Access

In this part of the assignment, you will create your tables in Access. Use the following guidelines:

- Enter customer records into the tables. Create eight customer records, including one for yourself.
- Enter as many songs as your instructor requires. To make this easier, you could import your own music library into the database. Make sure that the genre and song length are included in the imported or created data.

- For analysis, you'll only look at your own playlist, so create one that includes at least 40 songs.
- Appropriately limit the size of the text fields; for example, a telephone number does not need the default length of 255 characters.
- Print all tables, if required.

Assignment 2B: Creating Queries

You will create three queries, one of which will be exported to Excel with a table, as outlined in the Background section of this case.

Query 1

Create a query called Current Library that lists songs in the company's library and includes headings for Genre and Number of Songs. Order the genres in the query from most songs to least. Note the column heading change from the default setting provided by the query generator. Your query should resemble that in Figure 12-2, although the data will be different.

Current Library

Genre	Number of Songs
Alternative & Punk	478
Rock	428
Hip Hop/Rap	164
Alternative	153
Pop	83
Rap	62
Electronic	48
Rock/Pop	47
Hip-Hop	40
Soundtrack	40
Indie	39

FIGURE 12-2 Current Library query

Query 2

Create a query called What Artist? that prompts for an artist's name and lists songs by that artist. The query includes headings for Song Name, Song Time, and Artist. For example, if you entered "Led Zeppelin" at the prompt, the output might resemble that in Figure 12-3.

What Artist?

Song Name	Song Time	Artist
Kashmir	8:33	Led Zeppelin
Custard Pie	4:16	Led Zeppelin
The Rover	5:39	Led Zeppelin
In My Time Of Dying	11:08	Led Zeppelin
Houses Of The Holy	4:04	Led Zeppelin
Trampled Under Foot	5:39	Led Zeppelin
In The Light	8:49	Led Zeppelin
Bron-Yr-Aur	2:09	Led Zeppelin
Down By The Seaside	5:18	Led Zeppelin
Ten Years Gone	6:35	Led Zeppelin
Night Flight	3:39	Led Zeppelin
The Wanton Song	4:12	Led Zeppelin
Boogie With Stu	3:55	Led Zeppelin
Black Country Woman	4:35	Led Zeppelin
Sick Again	4:43	Led Zeppelin

FIGURE 12-3 What Artist? query

Query 3

Create a query called Listening Data that lists the last name of a customer, the playing time of a particular song, the song genre (using a numeric code), the number of times the customer has listened to the song, and the genre name. Create this query for the one customer you are analyzing. In this case, it's your playlist. Your data may differ, but the output should resemble that in Figure 12-4. Note that some column headings have been renamed for readability purposes.

Listening Data

Last Name	Song Time	Genre Code	Number of Times	Genre
Pao	4:51	2	5	Rock
Pao	3:50	2	2	Rock
Pao	5:02	2	3	Rock
Pao	3:42	2	6	Rock
Pao	3:58	2	1	Rock
Pao	4:20	2	1	Rock
Pao	2:35	2	10	Rock
Pao	3:23	2	9	Rock
Pao	2:38	2	1	Rock
Pao	2:33	2	3	Rock
Pao	3:00	2	4	Rock
Pao	3:09	2	1	Rock
Pao	2:27	2	12	Rock
Pao	2:25	1	1	Alternative & Punk
Pao	3:13	1	2	Alternative & Punk
Pao	3:02	1	3	Alternative & Punk
Pao	3:15	1	8	Alternative & Punk

FIGURE 12-4 Listening Data query

Assignment 2C: Importing Data and Doing Analysis

In this assignment, you import the data into Excel and analyze the data.

Importing the Data

Open Excel and save the file as Music.xlsx. Import the Listening Data query into a sheet in Excel. Rename the sheet Listening. Import the song table into a separate sheet and rename that sheet Songs. In each sheet, you'll have to reformat the numeric data to be numeric again (it imports as text). Note that each sheet's import should begin in cell A1.

Least Squares Analysis

In the Listening sheet, create the column headings shown in Figure 12-5.

G	H	I	J	K
Song Name	Song Time	Genre	Predicted Number of Times	Recommend?

FIGURE 12-5 Excel sheet column headings

Choose two songs from the Songs sheet. Copy the data from those songs into cells G2:I3. Use the Excel =trend function in cells J2 and J3 to predict the number of times you will listen to the songs. In cells K2 and K3, insert an =if statement that will judge whether the song should be recommended to the customer. If the function predicts that a customer will listen to a song more than seven times, recommend the song. Otherwise, do not recommend it. Note that you are attempting to predict enjoyment of a song based on its genre and playing time. Some people might like short, catchy tunes, and others might like long rock ballads.

Assume that you will include many songs in these columns later (e.g., G2:I500), so make sure that you use Excel addressing and referencing properly. Your analysis will yield results that resemble those in Figure 12-6.

G	H	I	J	K
Song Name	Song Time	Genre	Predicted Number of Times	Recommend?
Mr Brightside	0.16	1	6.970531226	Don't Recommend
In Da Pub	0.15	9	9.24032763	Recommend

FIGURE 12-6 Analysis and recommendations in Excel

ASSIGNMENT 3: MAKING A PRESENTATION

Create a presentation for the company that includes information about your database design. Demonstrate how your prediction works and suggest further predictive indicators such as musical attributes of songs. Include time for discussing how you will expand your analysis to include all customers' data. Your presentation should take fewer than 15 minutes, including a brief question-and-answer period.

DELIVERABLES

Assemble the following deliverables for your instructor, either electronically or in printed form:

1. Word-processed design of tables
2. Tables created in Access
3. Query 1: Current Library
4. Query 2: What Artist?
5. Query 3: Listening Data
6. Excel spreadsheet
7. Presentation materials
8. Any other required tutorial printouts or electronic media

Staple all pages together. Put your name and class number at the top of the page. Make sure that your electronic media are labeled, if required.

PART 6

ADVANCED EXCEL SKILLS

GUIDANCE FOR EXCEL CASES

The Excel cases in this book require the student to write a memorandum that includes a table. Guidelines for preparing a memorandum in Microsoft Word and instructions for entering a table in a Word document are listed to begin this tutorial. Also, some of the cases in this casebook require the use of advanced Excel techniques. Those techniques are explained in this tutorial rather than in the cases themselves:

- Using data tables
- Using pivot tables
- Using built-in functions

You can refer to Sheet 1 of TutEData.xlsx when reading about data tables. Refer to Sheet 2 when reading about pivot tables.

PREPARING A MEMORANDUM IN WORD

A business memorandum should include proper headings, such as TO/FROM /DATE/SUBJECT. If you want to use a Word memorandum template, follow these steps:

1. In Microsoft Word, click the Office button.
2. Click New.
3. Click the Memos template in the Templates section.
4. Choose the contemporary design memo.
5. Click OK.

The first time you do this, you may need to click Download to install the template.

ENTERING A TABLE INTO A WORD DOCUMENT

Enter a table into a Word document using the following procedure:

1. Click the cursor where you want the table to appear in the document.
2. In the Insert group, select the Table drop-down menu.
3. Select Insert Table.
4. Choose the number of rows and columns.
5. Click OK.

DATA TABLES

An Excel data table is a contiguous range of data that has been designated as a table. Once you make this designation, the table gains certain properties that are useful for data analysis. (*Note:* In previous versions of Excel, data tables were called *data lists*.) Suppose you have a list of runners who have run a race, as shown in Figure E-1.

	A	B	C	D	E	F
1	RUNNER#	LAST	FIRST	AGE	GENDER	TIME (MIN)
2	100	HARRIS	JANE	O	F	70
3	101	HILL	GLENN	Y	M	70
4	102	GARCIA	PEDRO	M	M	85
5	103	HILBERT	DORIS	M	F	90
6	104	DOAKS	SALLY	Y	F	94
7	105	JONES	SUE	Y	F	95
8	106	SMITH	PETE	M	M	100
9	107	DOE	JANE	O	F	100
10	108	BRADY	PETE	O	M	100
11	109	BRADY	JOE	O	M	120
12	110	HEEBER	SALLY	M	F	125
13	111	DOLTZ	HAL	O	M	130
14	112	PEEBLES	AL	Y	M	63

FIGURE E-1 Data table example

To turn the information into a data table (list), you highlight the data range, including headers, and select the Insert tab. Then, in the Tables group, click Table. You see the Create Table window, as shown in Figure E-2.

A1 *fx* RUNNER#

	A	B	C	D	E	F
1	RUNNER#	LAST	FIRST	AGE	GENDER	TIME (MIN)
2	100	HARRIS	JANE	O	F	70
3	101	HILL	GLENN	Y	M	70
4	102	GARCI				85
5	103	HILBEF				90
6	104	DOAKS				94
7	105	JONES				95
8	106	SMITH				100
9	107	DOE				100
10	108	BRADY				100
11	109	BRADY	JOE	O	M	120
12	110	HEEBER	SALLY	M	F	125
13	111	DOLTZ	HAL	O	M	130
14	112	PEEBLES	AL	Y	M	63

Create Table
Where is the data for your table?
=$A$1:$F$14
My table has headers
OK Cancel

FIGURE E-2 Create Table window

When you click OK, the data range appears as a table. In the Design tab, select Total Row to add a totals row to the data table. You also can select Table Styles – Light to get rid of the contrasting color in the table's rows. Figure E-3 shows the results.

	A	B	C	D	E	F
1	RUNNER#	LAST	FIRST	AGE	GENDER	TIME (MIN)
2	100	HARRIS	JANE	O	F	70
3	101	HILL	GLENN	Y	M	70
4	102	GARCIA	PEDRO	M	M	85
5	103	HILBERT	DORIS	M	F	90
6	104	DOAKS	SALLY	Y	F	94
7	105	JONES	SUE	Y	F	95
8	106	SMITH	PETE	M	M	100
9	107	DOE	JANE	O	F	100
10	108	BRADY	PETE	O	M	100
11	109	BRADY	JOE	O	M	120
12	110	HEEBER	SALLY	M	F	125
13	111	DOLTZ	HAL	O	M	130
14	112	PEEBLES	AL	Y	M	63
15	Total					1242

FIGURE E-3 Data table example

The headers have acquired drop-down menu tabs, as you can see in Figure E-3.

You can sort the data table records by any field. Perhaps you want to sort by Times. If so, click the drop-down menu in the TIME (MIN) header and select Sort – Smallest to Largest. You get the results shown in Figure E-4.

	A	B	C	D	E	F
1	RUNNER	LAST	FIRST	AGE	GENDE	TIME (MIN
2	112	PEEBLES	AL	Y	M	63
3	100	HARRIS	JANE	O	F	70
4	101	HILL	GLENN	Y	M	70
5	102	GARCIA	PEDRO	M	M	85
6	103	HILBERT	DORIS	M	F	90
7	104	DOAKS	SALLY	Y	F	94
8	105	JONES	SUE	Y	F	95
9	106	SMITH	PETE	M	M	100
10	107	DOE	JANE	O	F	100
11	108	BRADY	PETE	O	M	100
12	109	BRADY	JOE	O	M	120
13	110	HEEBER	SALLY	M	F	125
14	111	DOLTZ	HAL	O	M	130
15	Total					1242

FIGURE E-4 Sorting list by drop-down menu

You can see that Peebles had the best time and Doltz had the worst time. You also can sort from Largest to Smallest.

In addition, you can sort by more than one criterion. Assume you want to sort first by Gender and then by Time (within Gender). You first sort from Smallest to Largest in Gender. Then you again activate the Gender drop-down tab and select Sort By Color – Custom Sort. In the Sort window that appears, click Add Level and choose Time as the next criterion. See Figure E-5.

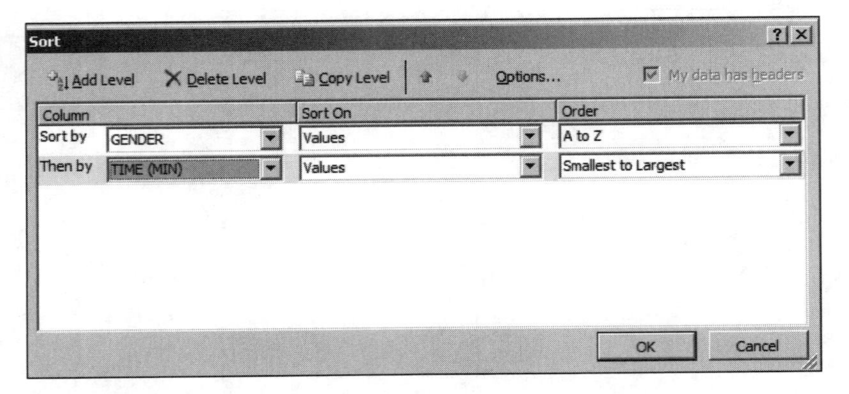

FIGURE E-5 Sorting on multiple criteria

Click OK to get the results shown in Figure E-6.

	A	B	C	D	E	F
1	RUNNER	LAST	FIRST	AGE	GENDE	TIME (MIN
2	100	HARRIS	JANE	O	F	70
3	103	HILBERT	DORIS	M	F	90
4	104	DOAKS	SALLY	Y	F	94
5	105	JONES	SUE	Y	F	95
6	107	DOE	JANE	O	F	100
7	110	HEEBER	SALLY	M	F	125
8	112	PEEBLES	AL	Y	M	63
9	101	HILL	GLENN	Y	M	70
10	102	GARCIA	PEDRO	M	M	85
11	106	SMITH	PETE	M	M	100
12	108	BRADY	PETE	O	M	100
13	109	BRADY	JOE	O	M	120
14	111	DOLTZ	HAL	O	M	130
15	Total					1242

FIGURE E-6 Sorting by Gender and Time (within Gender)

You can see that Harris had the best female time and that Peebles had the best male time.

Perhaps you want to see the Top *n* listings for some attribute; for example, you may want to see the top five runners' times. Select the Time column's drop-down menu and select Number Filters. You get another menu, in which you select Top 10. The Top 10 AutoFilter window appears, as shown in Figure E-7.

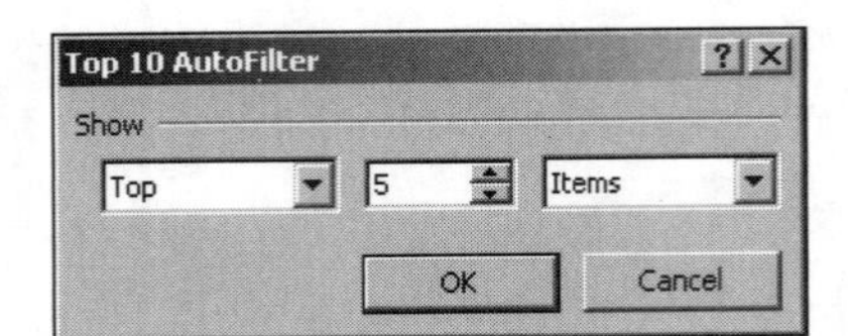

FIGURE E-7 Top 10 AutoFilter window

This window lets you specify the number of values you want. In Figure E-7, five values were specified. Click OK to get the results shown in Figure E-8.

	A	B	C	D	E	F
1	RUNNER	LAST	FIRST	AGE	GENDE	TIME (MIN
6	107	DOE	JANE	O	F	100
7	110	HEEBER	SALLY	M	F	125
11	106	SMITH	PETE	M	M	100
12	108	BRADY	PETE	O	M	100
13	109	BRADY	JOE	O	M	120
14	111	DOLTZ	HAL	O	M	130
15	Total					675

FIGURE E-8 Top 5 times

The output contains more than five data records because there are ties at 100 minutes. If you want to see all of the records again, click the Time drop-down menu and select Clear Filter. The full table of data reappears, as shown in Figure E-9.

	A	B	C	D	E	F
1	RUNNER	LAST	FIRST	AGE	GENDE	TIME (MIN
2	100	HARRIS	JANE	O	F	70
3	103	HILBERT	DORIS	M	F	90
4	104	DOAKS	SALLY	Y	F	94
5	105	JONES	SUE	Y	F	95
6	107	DOE	JANE	O	F	100
7	110	HEEBER	SALLY	M	F	125
8	112	PEEBLES	AL	Y	M	63
9	101	HILL	GLENN	Y	M	70
10	102	GARCIA	PEDRO	M	M	85
11	106	SMITH	PETE	M	M	100
12	108	BRADY	PETE	O	M	100
13	109	BRADY	JOE	O	M	120
14	111	DOLTZ	HAL	O	M	130
15	Total					1242

FIGURE E-9 Restoring all data to screen

Each of the cells in the Total row has a drop-down menu. The menu choices are statistical operations that you can perform on the totals—for example, you can take a sum, take an average, take a minimum or maximum, count the number of records, and so on. Assume the Time drop-down menu was selected, as shown in Figure E-10. Note that the Sum operator is highlighted by default.

	A	B	C	D	E	F
1	RUNNER	LAST	FIRST	AGE	GENDE	TIME (MIN
2	100	HARRIS	JANE	O	F	70
3	103	HILBERT	DORIS	M	F	90
4	104	DOAKS	SALLY	Y	F	94
5	105	JONES	SUE	Y	F	95
6	107	DOE	JANE	O	F	100
7	110	HEEBER	SALLY	M	F	125
8	112	PEEBLES	AL	Y	M	63
9	101	HILL	GLENN	Y	M	70
10	102	GARCIA	PEDRO	M	M	85
11	106	SMITH	PETE	M	M	100
12	108	BRADY	PETE	O	M	100
13	109	BRADY	JOE	O	M	120
14	111	DOLTZ	HAL	O	M	130
15	Total					1242
16						None
17						Average
18						Count
19						Count Numbers
20						Max
21						Min
						Sum
						StdDev
						Var
						More Functions...

FIGURE E-10 Selecting Time drop-down menu in Total row

By changing from Sum to the Average operator, you find that the average time for all runners was 95.5 minutes, as shown in Figure E-11.

	A	B	C	D	E	F
1	RUNNER	LAST	FIRST	AGE	GENDE	TIME (MIN
2	100	HARRIS	JANE	O	F	70
3	103	HILBERT	DORIS	M	F	90
4	104	DOAKS	SALLY	Y	F	94
5	105	JONES	SUE	Y	F	95
6	107	DOE	JANE	O	F	100
7	110	HEEBER	SALLY	M	F	125
8	112	PEEBLES	AL	Y	M	63
9	101	HILL	GLENN	Y	M	70
10	102	GARCIA	PEDRO	M	M	85
11	106	SMITH	PETE	M	M	100
12	108	BRADY	PETE	O	M	100
13	109	BRADY	JOE	O	M	120
14	111	DOLTZ	HAL	O	M	130
15	Total					95.53846154

FIGURE E-11 Average running time shown in Total row

PIVOT TABLES

Suppose you have data for a company's sales transactions by month, by salesperson, and by amount for each product type. You would like to display each salesperson's total sales by type of product sold and by month. You can use a pivot table in Excel to tabulate that summary data. A pivot table is built around one or more dimensions and thus can summarize large amounts of data.

Figure E-12 shows total sales cross-tabulated by salesperson and by month. The following steps explain how to create a pivot chart from the data.

	A	B	C	D	E
1	Name	Product	January	February	March
2	Jones	Product1	30,000	35,000	40,000
3	Jones	Product2	33,000	34,000	45,000
4	Jones	Product3	24,000	30,000	42,000
5	Smith	Product1	40,000	38,000	36,000
6	Smith	Product2	41,000	37,000	38,000
7	Smith	Product3	39,000	50,000	33,000
8	Bonds	Product1	25,000	26,000	25,000
9	Bonds	Product2	22,000	25,000	24,000
10	Bonds	Product3	19,000	20,000	19,000
11	Ruth	Product1	44,000	42,000	33,000
12	Ruth	Product2	45,000	40,000	30,000
13	Ruth	Product3	50,000	52,000	35,000

FIGURE E-12 Excel spreadsheet data

You can create pivot tables and many other kinds of tables with the Excel PivotTable tool. To create a pivot table from the data in Figure E-12, follow these steps:

1. Starting in the spreadsheet in Figure E-12, go to the Insert tab. In the Tables group, choose PivotTable. You see the screen shown in Figure E-13.

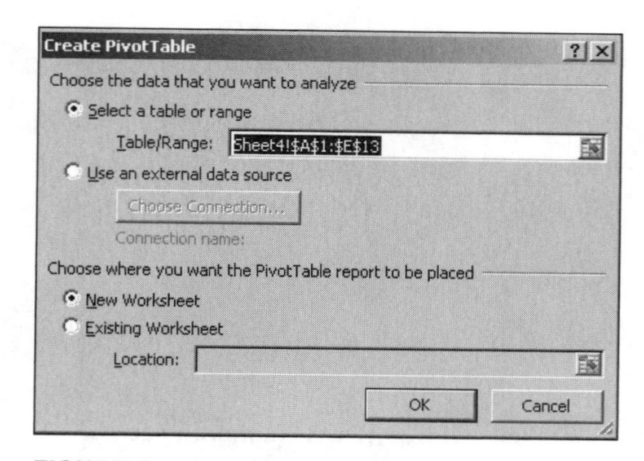

FIGURE E-13 Creating a pivot table

2. Make sure New Worksheet is checked under "Choose where you want the PivotTable report to be placed." Click OK. The screen shown in Figure E-14 appears. If it does not, right-click in a cell in the pivot table area. Select Pivot Table Options from the menu. Click the Display tab and then check the Classic Layout.

FIGURE E-14 PivotTable design screen

The data range's column headings are shown in the PivotTable Field List on the right side of the screen. From there, you can click and drag column headings into the Row, Column, and Data areas that appear in the spreadsheet.

3. If you want to see the total sales by product for each salesperson, drag the Name field to the Drop Column Fields Here area in the spreadsheet. You should see the result shown in Figure E-15.

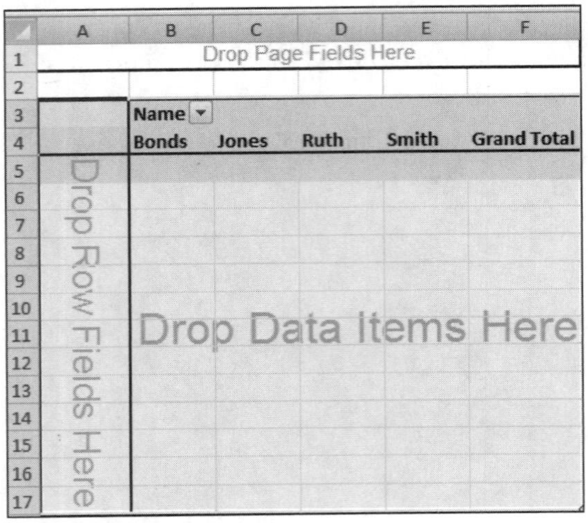

FIGURE E-15 Column fields

4. Next, take the Product field and drag it to the Drop Row Fields Here area. You should see the result shown in Figure E-16.

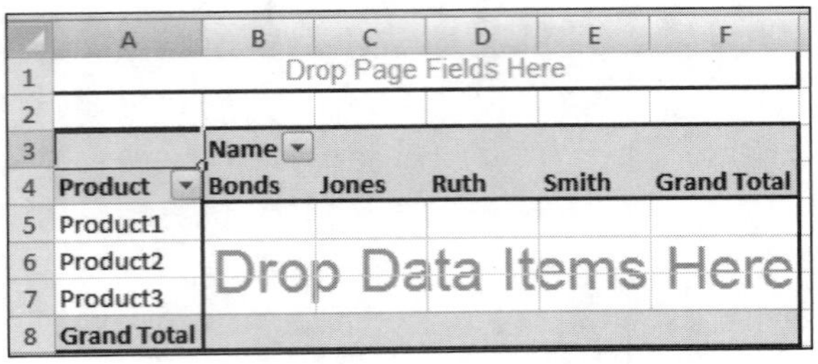

FIGURE E-16 Row fields

5. Finally, take the month fields (January, February, and March) and drag them individually to the Drop Data Items Here area to produce the finalized pivot table. You should see the result shown in Figure E-17.

	A	B	C	D	E	F	G
1	Drop Page Fields Here						
2							
3			Name				
4	Product	Values	Bonds	Jones	Ruth	Smith	Grand Total
5	Product1	Sum of January	25000	30000	44000	40000	139000
6		Sum of February	26000	35000	42000	38000	141000
7		Sum of March	25000	40000	33000	36000	134000
8	Product2	Sum of January	22000	33000	45000	41000	141000
9		Sum of February	25000	34000	40000	37000	136000
10		Sum of March	24000	45000	30000	38000	137000
11	Product3	Sum of January	19000	24000	50000	39000	132000
12		Sum of February	20000	30000	52000	50000	152000
13		Sum of March	19000	42000	35000	33000	129000
14	Total Sum of January		66000	87000	139000	120000	412000
15	Total Sum of February		71000	99000	134000	125000	429000
16	Total Sum of March		68000	127000	98000	107000	400000

FIGURE E-17 Data items

By default, Excel adds all of the sales for each salesperson by month for each product. At the bottom of the pivot table, Excel also shows the total sales for each month for all products.

BUILT-IN FUNCTIONS

The following functions are referred to in the Excel cases in this text:

MIN, MAX, AVERAGE, COUNTIF, and ROUND

The syntax of these functions is discussed here. The following examples are based on the runner data shown in Figure E-18.

	A	B	C	D	E	F	G
1	RUNNER#	LAST	FIRST	AGE	GENDER	HEIGHT	TIME (MIN)
2	100	HARRIS	JANE	O	F	60	70
3	101	HILL	GLENN	Y	M	65	70
4	102	GARCIA	PEDRO	M	M	76	85
5	103	HILBERT	DORIS	M	F	64	90
6	104	DOAKS	SALLY	Y	F	62	94
7	105	JONES	SUE	Y	F	64	95
8	106	SMITH	PETE	M	M	73	100
9	107	DOE	JANE	O	F	66	100
10	108	BRADY	PETE	O	M	73	100
11	109	BRADY	JOE	O	M	71	120
12	110	HEEBER	SALLY	M	F	59	125
13	111	DOLTZ	HAL	O	M	76	130
14	112	PEEBLES	AL	Y	M	76	63

FIGURE E-18 Runner data used to illustrate built-in functions

Note that the data is the same as that shown in Figure E-1, except that Figure E-18 has a column for the runner's height in inches.

MIN and MAX Functions

The MIN function determines the smallest value in a range of data. The MAX function returns the largest. Say that we want to know the fastest time for all runners, which would be the minimum time in column G. The MIN function computes the smallest value in a set of values. The set of values could be a data range or it could be a series of cell addresses separated by commas. The syntax of the MIN function is as follows:

MIN(set of data)

To show the minimum time in cell C16, you would enter the formula shown in Figure E-19's formula bar.

C16 fx =MIN(G2:G14)

	A	B	C	D	E	F	G
1	RUNNER#	LAST	FIRST	AGE	GENDER	HEIGHT	TIME (MIN)
2	100	HARRIS	JANE	O	F	60	70
3	101	HILL	GLENN	Y	M	65	70
4	102	GARCIA	PEDRO	M	M	76	85
5	103	HILBERT	DORIS	M	F	64	90
6	104	DOAKS	SALLY	Y	F	62	94
7	105	JONES	SUE	Y	F	64	95
8	106	SMITH	PETE	M	M	73	100
9	107	DOE	JANE	O	F	66	100
10	108	BRADY	PETE	O	M	73	100
11	109	BRADY	JOE	O	M	71	120
12	110	HEEBER	SALLY	M	F	59	125
13	111	DOLTZ	HAL	O	M	76	130
14	112	PEEBLES	AL	Y	M	76	63
15							
16	MINIMUM TIME:		63				

FIGURE E-19 MIN function in cell C16

(Assume you typed the label "MINIMUM TIME:" into cell A16.) You can see that the fastest time is 63 minutes.

To see the slowest time in cell G16, use the MAX function, whose syntax parallels that of the MIN function, except that the largest value in the set is determined. See Figure E-20.

G16 fx =MAX(G2:G14)

	A	B	C	D	E	F	G
1	RUNNER#	LAST	FIRST	AGE	GENDER	HEIGHT	TIME (MIN)
2	100	HARRIS	JANE	O	F	60	70
3	101	HILL	GLENN	Y	M	65	70
4	102	GARCIA	PEDRO	M	M	76	85
5	103	HILBERT	DORIS	M	F	64	90
6	104	DOAKS	SALLY	Y	F	62	94
7	105	JONES	SUE	Y	F	64	95
8	106	SMITH	PETE	M	M	73	100
9	107	DOE	JANE	O	F	66	100
10	108	BRADY	PETE	O	M	73	100
11	109	BRADY	JOE	O	M	71	120
12	110	HEEBER	SALLY	M	F	59	125
13	111	DOLTZ	HAL	O	M	76	130
14	112	PEEBLES	AL	Y	M	76	63
15							
16	MINIMUM TIME:		63		MAXIMUM TIME:		130

FIGURE E-20 MAX function in cell G16

AVERAGE and ROUND Functions

The AVERAGE function computes the average of a set of values. Figure E-21 shows the use of the AVERAGE function in cell C-17.

C17 | fx =AVERAGE(G2:G14)

	A	B	C	D	E	F	G
1	RUNNER#	LAST	FIRST	AGE	GENDER	HEIGHT	TIME (MIN)
2	100	HARRIS	JANE	O	F	60	70
3	101	HILL	GLENN	Y	M	65	70
4	102	GARCIA	PEDRO	M	M	76	85
5	103	HILBERT	DORIS	M	F	64	90
6	104	DOAKS	SALLY	Y	F	62	94
7	105	JONES	SUE	Y	F	64	95
8	106	SMITH	PETE	M	M	73	100
9	107	DOE	JANE	O	F	66	100
10	108	BRADY	PETE	O	M	73	100
11	109	BRADY	JOE	O	M	71	120
12	110	HEEBER	SALLY	M	F	59	125
13	111	DOLTZ	HAL	O	M	76	130
14	112	PEEBLES	AL	Y	M	76	63
15							
16	MINIMUM TIME:		63		MAXIMUM TIME:		130
17	AVERAGE TIME:		95.53846				

FIGURE E-21 AVERAGE function in cell C17

Notice that the value shown is a real number with many digits. What if you wanted to have the value rounded to a certain number of digits? Of course, you could format the output cell, but doing that only changes what is shown on the screen. You want the cell's contents actually to *be* the rounded number. Therefore, you need to use the ROUND function. Its syntax is:

ROUND(number, number of digits)

Figure E-22 shows the rounded average time (2 decimals) in cell G17.

G17 | fx =ROUND(C17,2)

	A	B	C	D	E	F	G
1	RUNNER#	LAST	FIRST	AGE	GENDER	HEIGHT	TIME (MIN)
2	100	HARRIS	JANE	O	F	60	70
3	101	HILL	GLENN	Y	M	65	70
4	102	GARCIA	PEDRO	M	M	76	85
5	103	HILBERT	DORIS	M	F	64	90
6	104	DOAKS	SALLY	Y	F	62	94
7	105	JONES	SUE	Y	F	64	95
8	106	SMITH	PETE	M	M	73	100
9	107	DOE	JANE	O	F	66	100
10	108	BRADY	PETE	O	M	73	100
11	109	BRADY	JOE	O	M	71	120
12	110	HEEBER	SALLY	M	F	59	125
13	111	DOLTZ	HAL	O	M	76	130
14	112	PEEBLES	AL	Y	M	76	63
15							
16	MINIMUM TIME:		63		MAXIMUM TIME:		130
17	AVERAGE TIME:		95.53846		ROUNDED AVERAGE:		95.54

FIGURE E-22 ROUND function used in cell G17

To achieve this output, the cell C17 was used as the value to be rounded. Recall from Figure E-21 that cell C17 had the formula =AVERAGE(G2:G14). This ROUND formula would have given the same output in cell G17: =ROUND(AVERAGE(G2:G14),2). In this case, Excel evaluates the formula "inside out." First, the AVERAGE function is evaluated, yielding the average with the many digits. That value is then input to the ROUND function and rounded to two decimals.

COUNTIF Function

The COUNTIF function counts the number of values in a range that meet a specified condition. The syntax is:

COUNTIF(range of data, condition)

The condition is a logical expression such as "=1", ">6", or "=F". The condition is shown with quotation marks, even if a number is involved.

Assume that you want to see the number of female runners in cell C18. Figure E-23 shows the formula used.

C18 =COUNTIF(E2:E14,"F")

	A	B	C	D	E	F	G
1	RUNNER#	LAST	FIRST	AGE	GENDER	HEIGHT	TIME (MIN)
2	100	HARRIS	JANE	O	F	60	70
3	101	HILL	GLENN	Y	M	65	70
4	102	GARCIA	PEDRO	M	M	76	85
5	103	HILBERT	DORIS	M	F	64	90
6	104	DOAKS	SALLY	Y	F	62	94
7	105	JONES	SUE	Y	F	64	95
8	106	SMITH	PETE	M	M	73	100
9	107	DOE	JANE	O	F	66	100
10	108	BRADY	PETE	O	M	73	100
11	109	BRADY	JOE	O	M	71	120
12	110	HEEBER	SALLY	M	F	59	125
13	111	DOLTZ	HAL	O	M	76	130
14	112	PEEBLES	AL	Y	M	76	63
15							
16	MINIMUM TIME:		63		MAXIMUM TIME:		130
17	AVERAGE TIME:		95.53846		ROUNDED AVERAGE:		95.54
18	NUMBER OF FEMALES:		6				

FIGURE E-23 COUNTIF function used in cell C18

The logic of the formula is: Count the number of times that "F" appears in the data range E2:E14.

As another example of using COUNTIF, assume that column H shows the rounded ratio of the runner's height in inches to the runner's time in minutes (see Figure E-24).

H2 =ROUND(G2/F2,2)

	A	B	C	D	E	F	G	H
1	RUNNER#	LAST	FIRST	AGE	GENDER	HEIGHT	TIME (MIN)	RATIO
2	100	HARRIS	JANE	O	F	60	70	1.17
3	101	HILL	GLENN	Y	M	65	70	1.08
4	102	GARCIA	PEDRO	M	M	76	85	1.12
5	103	HILBERT	DORIS	M	F	64	90	1.41
6	104	DOAKS	SALLY	Y	F	62	94	1.52
7	105	JONES	SUE	Y	F	64	95	1.48
8	106	SMITH	PETE	M	M	73	100	1.37
9	107	DOE	JANE	O	F	66	100	1.52
10	108	BRADY	PETE	O	M	73	100	1.37
11	109	BRADY	JOE	O	M	71	120	1.69
12	110	HEEBER	SALLY	M	F	59	125	2.12
13	111	DOLTZ	HAL	O	M	76	130	1.71
14	112	PEEBLES	AL	Y	M	76	63	0.83
15								
16	MINIMUM TIME:		63		MAXIMUM TIME:		130	
17	AVERAGE TIME:		95.53846		ROUNDED AVERAGE:		95.54	
18	NUMBER OF FEMALES:		6					

FIGURE E-24 Ratio of height to time in column H

Assume that all runners whose height in inches is less than their time in minutes will get an award. How many awards are needed? If the ratio is less than 1, an award is warranted. The COUNTIF function in cell G18 computes a count of ratios less than 1, as shown in Figure E-25.

G18 fx =COUNTIF(H2:H14,"<1")

	A	B	C	D	E	F	G	H
1	RUNNER#	LAST	FIRST	AGE	GENDER	HEIGHT	TIME (MIN)	RATIO
2	100	HARRIS	JANE	O	F	60	70	1.17
3	101	HILL	GLENN	Y	M	65	70	1.08
4	102	GARCIA	PEDRO	M	M	76	85	1.12
5	103	HILBERT	DORIS	M	F	64	90	1.41
6	104	DOAKS	SALLY	Y	F	62	94	1.52
7	105	JONES	SUE	Y	F	64	95	1.48
8	106	SMITH	PETE	M	M	73	100	1.37
9	107	DOE	JANE	O	F	66	100	1.52
10	108	BRADY	PETE	O	M	73	100	1.37
11	109	BRADY	JOE	O	M	71	120	1.69
12	110	HEEBER	SALLY	M	F	59	125	2.12
13	111	DOLTZ	HAL	O	M	76	130	1.71
14	112	PEEBLES	AL	Y	M	76	63	0.83
15								
16	MINIMUM TIME:		63		MAXIMUM TIME:		130	
17	AVERAGE TIME:		95.53846		ROUNDED AVERAGE:		95.54	
18	NUMBER OF FEMALES:		6		RATIOS<1:		1	

FIGURE E-25 COUNTIF function used in cell G18

PART 7

PRESENTATION SKILLS

GIVING AN ORAL PRESENTATION

Giving an oral presentation provides you the opportunity to practice the presentation skills you'll need in the workplace. The presentations you create for the cases in this textbook will be similar to real-world presentations. You'll present objective, technical results to an organization's stakeholders, and you'll support your presentation with visual aids commonly used in the business world. During your presentation, your instructor might assign your classmates to role-play an audience of business managers, bankers, or employees and ask them to give you feedback on your presentation.

Follow these four steps to create an effective presentation:

1. Plan your presentation.
2. Draft your presentation.
3. Create graphics and other visual aids.
4. Practice your delivery.

You'll start at the beginning and look at the steps involved in planning your presentation.

PLANNING YOUR PRESENTATION

When planning an oral presentation, you need to be aware of your time limits, establish your purpose, analyze your audience, and gather information. This section will look at each of those elements.

Knowing Your Time Limits

You need to consider your time limits on two levels. First, consider how much time you'll have to deliver your presentation. For example, what can you expect to accomplish in ten minutes? The element of time is the driver of any presentation. It limits the breadth and depth of your talk—and the number of visual aids that you can use. Second, consider how much time you'll need for the actual process of preparing your presentation—that is, for drafting your presentation, creating graphics, and practicing your delivery.

Establishing Your Purpose

After considering your time limits, you must define your purpose: what you need and want to say and to whom you will say it. For the cases in the Access portion of this book, your purpose will be to inform and explain. For instance, a business's owners, managers, and employees may need to know how their organization's database is organized and how they could use it to fill in input forms and create reports. In contrast, for the cases in the Excel portion of the book, your purpose will be to recommend a course of action. You'll be making recommendations to business owners, managers, and bankers based on the results you obtained from inputting and running various scenarios.

Analyzing Your Audience

Once you have established the purpose of your presentation, you should analyze your audience. Ask yourself these questions: What does my audience already know about the subject? What do the audience members want to know? What do they need to know? Do they have any biases that I should consider? What level of technical detail is best suited to their level of knowledge and interest?

In some Access cases, you will make a presentation to an audience that might not be familiar with Access or with databases in general. In other cases, you might be giving a presentation to a business owner who started to work on the database but was not able to finish it. Tailor your presentation to suit your audience.

For the Excel cases, you will be interpreting results for an audience of bankers and business managers. The audience will not need to know the detailed technical aspects of how you generated your results.

However, those listeners will need to know what assumptions you made prior to developing your spreadsheet, because those assumptions might have an impact on their opinion of your results.

Gathering Information

Because you will have just completed a case as you begin preparing your oral presentation, you'll have the basic information you need. For the Access cases, you should review the main points of the case and your goals. Make sure you include all of the points you think are important for the audience to understand. In addition, you might want to go beyond the requirements and explain additional ways in which the database could be used to benefit the organization, now or in the future.

For the Excel cases, you can refer to the tutorials for assistance in interpreting the results from your spreadsheet analysis. For some cases, you might want to research the Internet for business trends or background information that you can use to support your presentation.

DRAFTING YOUR PRESENTATION

Now that you have completed the planning stage, you are ready to begin drafting your presentation. At this point, you might be tempted to write your presentation and then memorize it word for word. If you do, your presentation will sound unnatural, because when people speak, they use a simpler vocabulary and shorter sentences than when they write. Thus, you might consider drafting your presentation by simply noting key phrases and statistics. When drafting your presentation, follow this sequence:

1. Write the main body of your presentation.
2. Write the introduction to your presentation.
3. Write the conclusion to your presentation.

Writing the Main Body

When you draft your presentation, write the body first. If you try to write the opening paragraph first, you'll spend an inordinate amount of time creating a "perfect" paragraph—only to revise it after you've written the body of your presentation.

Keeping Your Audience in Mind

To write the main body, review your purpose and your audience's profile. What are the main points you need to make? What are your audience's wants, needs, interests, and technical expertise? It's important to include some basic technical details in your presentation, but keep in mind the technical expertise of your audience.

What if your audience consists of people with varying needs, interests, and levels of expertise? For example, in the Access cases, an employee might want to know how to input information into a form, but the business owner might already know how to input data and therefore be more interested in learning how to generate queries and reports. You'll need to acknowledge their differences in your presentation. For example, you might say, "And now, let's look at how data entry clerks can input data into the form."

Similarly, in the Excel cases, your audience will usually consist of business owners, managers, and bankers. The owners' and managers' concerns will be profitability and growth. In contrast, the bankers' main concern will be repayment of a loan. You'll need to address the interests of each group.

Using Transitions and Repetition

Because your audience can't read the text of your presentation, you'll need to use transitions to compensate. Words such as *next*, *first*, *second*, and *finally* will help your audience follow the sequence of your ideas. Words such as *however*, *in contrast*, *on the other hand*, and *similarly* will help the audience follow shifts in thought. You also can use your voice and hand gestures to convey emphasis.

Also think about how you can use body language to emphasize what you're saying. For instance, if you are stating three reasons, you can use your fingers to tick off each reason as you discuss it: one, two, three. Similarly, if you're saying that profits will be flat, you can make a level motion with your hand for emphasis.

As you draft your presentation, repeat key points to emphasize them. For example, suppose your point is that outsourcing labor will provide the greatest gains in net income. Begin by previewing that concept. State that you're going to demonstrate how outsourcing labor will yield the biggest profits. Then provide statistics that support your claim, and show visual aids that graphically illustrate your point. Summarize by repeating your point: "As you can see, outsourcing labor does yield the biggest profits."

Relying on Graphics to Support Your Talk

As you write the main body, think of how you can best incorporate graphics into your presentation. Don't waste words describing what you're presenting, when you can use a graphic to portray the subject quickly. For instance, instead of describing how information from a query is input into a report, show a sample, a query result, and a completed report. Figure F-1 and Figure F-2 show an Access query and the resulting report, respectively.

Customer Name	Customer City	Product Name	Quantity	Total Value
Altamar	Miami	Xlarge	15	$825.00
Café Pacific	Dallas	Medium	10	$300.00
Café Pacific	Dallas	Large	10	$420.00
Café Pacific	Dallas	Xlarge	10	$550.00
Red Lobster	Wichita	Xlarge	15	$825.00
Zee Grill	Toronto	Large	25	$1,050.00
Big Fish	Dearborn	Medium	5	$150.00
Big Fish	Dearborn	Large	5	$210.00
Big Fish	Dearborn	Xlarge	5	$275.00
The Park Hotel	Charlotte	Medium	3	$90.00
St. Elmo Steak House	Indianapolis	Large	10	$420.00
Bound'Ry Restaurant	Nashville	Xlarge	15	$825.00
Bellagio	Las Vegas	Large	5	$210.00
Osetra The Fish House	San Diego	Medium	12	$360.00
*				

FIGURE F-1 Access query

Weekly Sales

Customer Name	*Customer City*	*Product Name*	*Quantity*	*Total Value*
Altamar	Miami			
		Xlarge	15	$825.00
Total Quantity and Value			*15*	*$825.00*
Bellagio	Las Vegas			
		Large	5	$210.00
Total Quantity and Value			*5*	*$210.00*
Big Fish	Dearborn			
		Xlarge	5	$275.00
		Large	5	$210.00
		Medium	5	$150.00
Total Quantity and Value			*15*	*$635.00*
Bound'Ry Restaurant	Nashville			
		Xlarge	15	$825.00
Total Quantity and Value			*15*	*$825.00*
Café Pacific	Dallas			
		Xlarge	10	$550.00
		Large	10	$420.00
		Medium	10	$300.00
Total Quantity and Value			*30*	*$1,270.00*
Osetra The Fish House	San Diego			

FIGURE F-2 Access report

Also consider what kinds of graphics media are available—and how well you can use them. For example, if you've never used Microsoft PowerPoint to prepare a presentation, will you have enough time to learn the software before you deliver your upcoming presentation? (If you don't know how to use PowerPoint, you might consider finding a tutorial on the Web to help you learn the basics.)

Anticipating the Unexpected

Even though you're just drafting your presentation at this stage, eventually you'll be answering questions from the audience. Being able to handle questions smoothly is the mark of a professional. The first steps to addressing audience questions are being able to anticipate them and preparing your answers.

You won't use all of the facts you gather for your presentation. However, as you draft your presentation, you might want to jot down some of those facts and keep them handy—just in case you need them to answer questions from the audience. For instance, during some Excel presentations, you might be asked why you are not recommending a certain course of action or why you did not mention that course of action in your report.

Writing the Introduction

After you have written the main body of your talk, you will want to develop an introduction. An introduction should be only a paragraph or two in length and should preview the main points that your presentation will cover.

For some of the Access cases, you might want to include general information about databases: what they can do, why they are used, and how they can help the company become more efficient and profitable. You won't need to say much about the business operation because the audience already works for the company.

For the Excel cases, you might want to include an introduction to the general business scenario and describe any assumptions you made when creating and running your decision support spreadsheet. Excel is used for decision support, so you should describe the choices and decision criteria you faced.

Writing the Conclusion

Every good presentation needs a good ending. Don't leave the audience hanging. Your conclusion should be brief—only a paragraph or two—and it should give your presentation a sense of closure. Use the conclusion to repeat your main points or, for the Excel cases, to recap your findings and/or recommendations.

CREATING GRAPHICS

Using visual aids is a powerful means of getting your point across and making it understandable to your audience. Visual aids come in a variety of forms, some of which are more effective than others.

Choosing Graphics Media

The media you use should depend on your situation and the media you have available. One of the key points to remember when using any media is this: *You must maintain control of the media, or you'll lose control of your audience.*

The following list highlights some of the most common media and their strengths and weaknesses:

- **Handouts:** This medium is readily available in classrooms and in businesses. It relieves the audience from taking notes. The graphics in handouts can be multicolored and of professional quality. *Negatives:* You must stop and take time to hand out individual pages. During your presentation, the audience may study and discuss your handouts rather than listen to you. Lack of media control is *the* major drawback—and it can hurt your presentation.
- **Chalkboard (or whiteboard):** This informal medium is readily available in the classroom, but not in many businesses. *Negatives:* You need to turn your back to the audience when you write (thereby running the risk of losing the audience's attention), and you need to erase what you've written as you continue your presentation. With chalkboards, your handwriting must be good—that is, it should be legible even when you write quickly. In addition, attractive graphics are difficult to create.
- **Flip Chart:** This informal medium is readily available in many businesses. *Negatives:* The writing space is so small that it's effective only for a very small audience. This medium shares many of the same negatives as the chalkboard.

- **Overheads:** This medium is readily available in classrooms and in businesses. You have control over what the audience sees and when they see it. You can create professional PowerPoint presentations on overhead transparencies. *Negatives:* If you don't use an application such as PowerPoint, handwritten overheads look amateurish. Without special equipment, professional-looking graphics are difficult to prepare.
- **Slides:** This formal medium is readily available in many businesses and can be used in large rooms. You can use 35-mm slides or the more popular electronic on-screen slides. In fact, electronic on-screen slides are usually *the* medium of choice for large organizations and are generally preferred for formal presentations. *Negatives:* You must have access to the equipment needed for slide presentations and know how to use it. It takes time to learn how to create and use computer graphics. Also, you must have some source of ambient light; otherwise, you may have difficulty seeing your notes in the dark.

Creating Charts and Graphs

Technically, charts and graphs are not the same thing, although many graphs are referred to as charts. Usually charts show relationships and graphs show change. However, Excel makes no distinction and calls both entities charts.

Charts are easy to create in Excel. Unfortunately, they are so easy to create that people often use graphics that are meaningless or that inaccurately reflect the data the graphics represent. Next, you'll look at how to select the most appropriate graphics.

You should use pie charts to display data that is related to a whole. For example, you might use a pie chart when showing the percentage of shoppers who bought a generic brand of toothpaste versus a major brand, as shown in Figure F-3. (Note that when creating a pie chart, Excel takes the numbers you want to graph and makes them a percentage of 100.) You would *not*, however, use a pie chart to show a company's net income over a three-year period. While Figure F-4 does show such a pie chart, the graphic is not meaningful because it is not useful to think of the period as "a whole" or the years as its "parts."

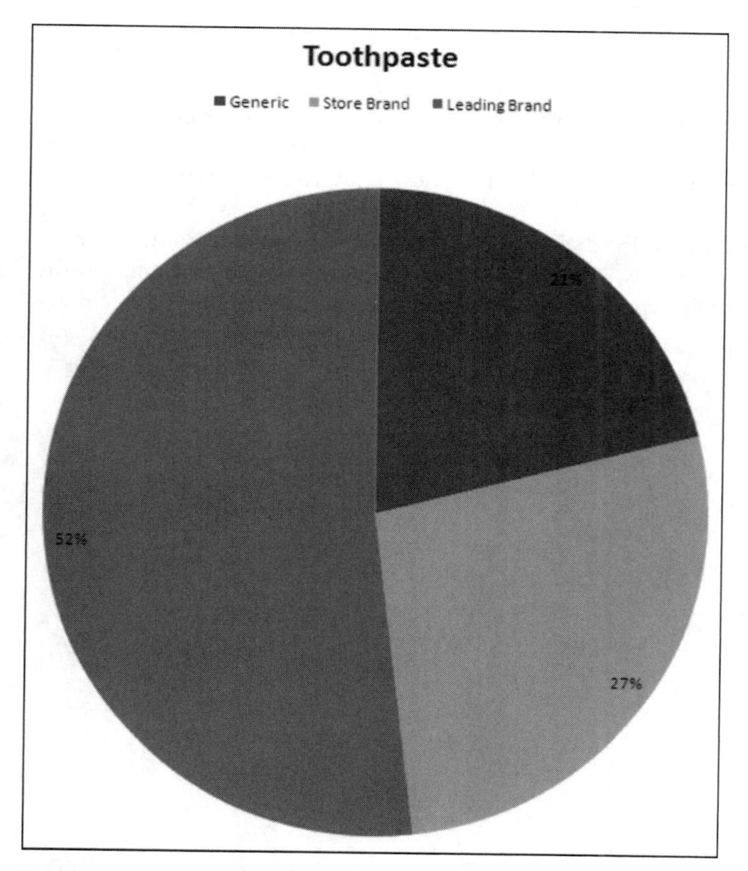

FIGURE F-3 Pie chart: appropriate use

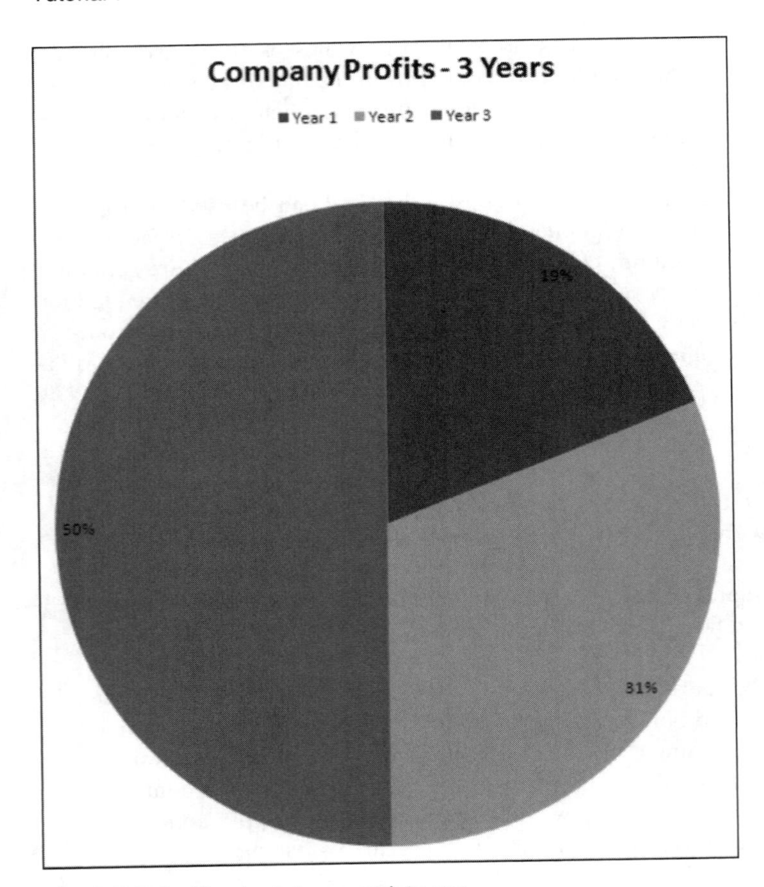

FIGURE F-4 Pie chart: inappropriate use

You should use bar charts when you want to compare several amounts at one time. For example, you might want to compare the net profit that would result from each of several different strategies. You also can use a bar chart to show changes over time. For example, you might show how one pricing strategy would increase profits year after year.

When you are showing a graphic, you need to include labels that explain what the graphic shows. For instance, when you're using a graph with an x- and y-axis, you should show what each axis represents, so the audience doesn't puzzle over the graphic while you're speaking. Figure F-5 and Figure F-6 show the necessity of labels.

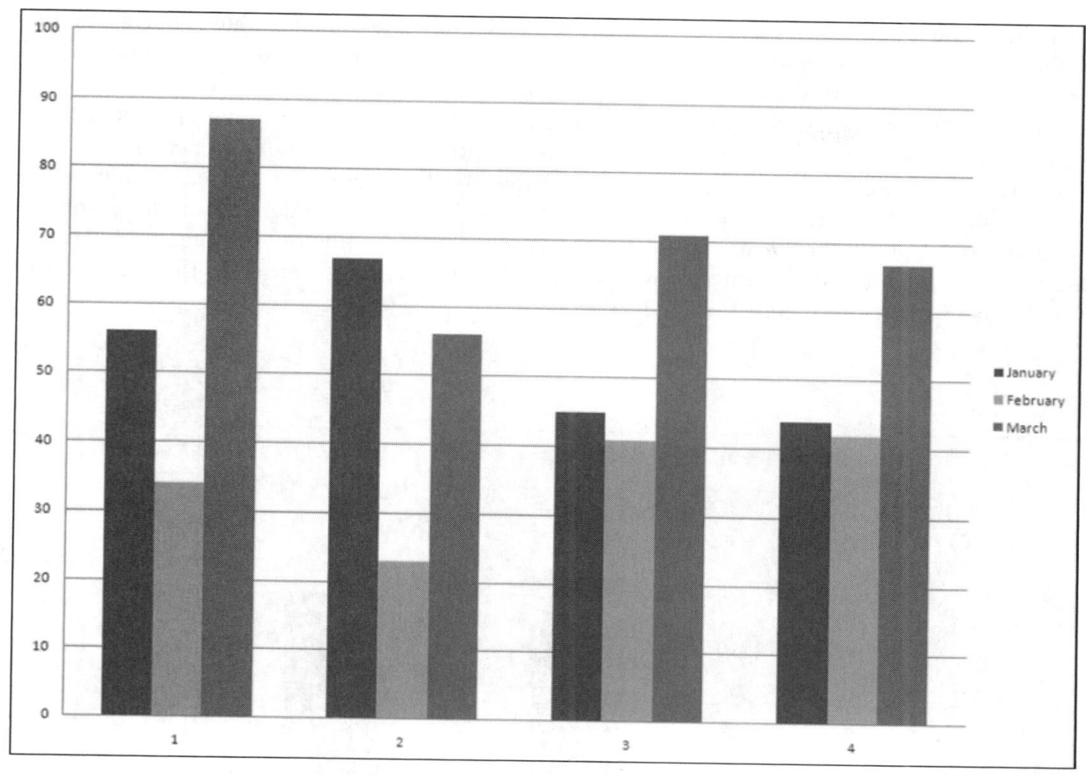

FIGURE F-5 Graphic without labels

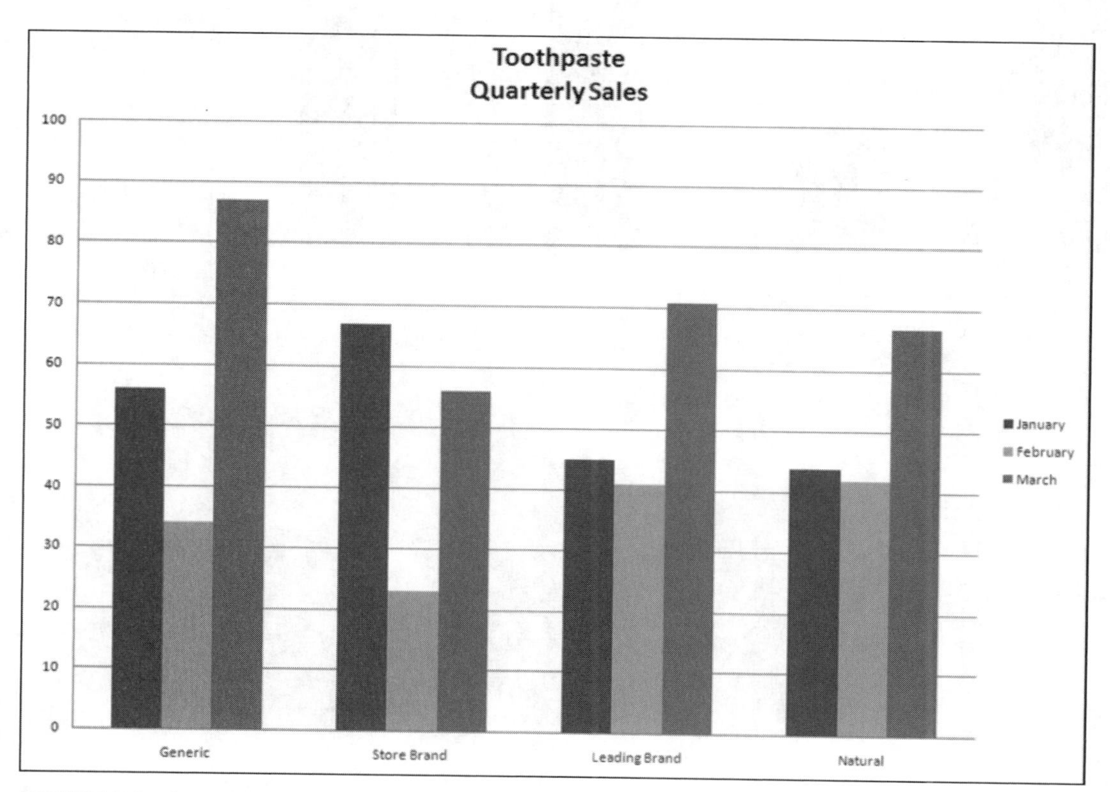

FIGURE F-6 Graphic with labels

In Figure F-5, neither the graphic nor the x- and y-axes are labeled. Do the amounts shown correspond to units or dollars? What elements are represented by each bar? In contrast, Figure F-6 provides a comprehensive snapshot of the business operation, which would support a talk rather than distract from it.

Another common pitfall of producing visual aids is creating charts that have a misleading premise. For example, suppose you want to show how sales have increased and contributed to a growth in net income. If you simply graph the number of items sold in a given month, as displayed in Figure F-7, the visual may not give your audience any sense of the actual dollar value of those items. Therefore, it might be more appropriate (and more revealing) to graph the profit margin for the items sold times the number of items sold. Graphing the profit margin would give a more accurate picture of which item(s) are contributing to the increased net income. That graph is displayed in Figure F-8.

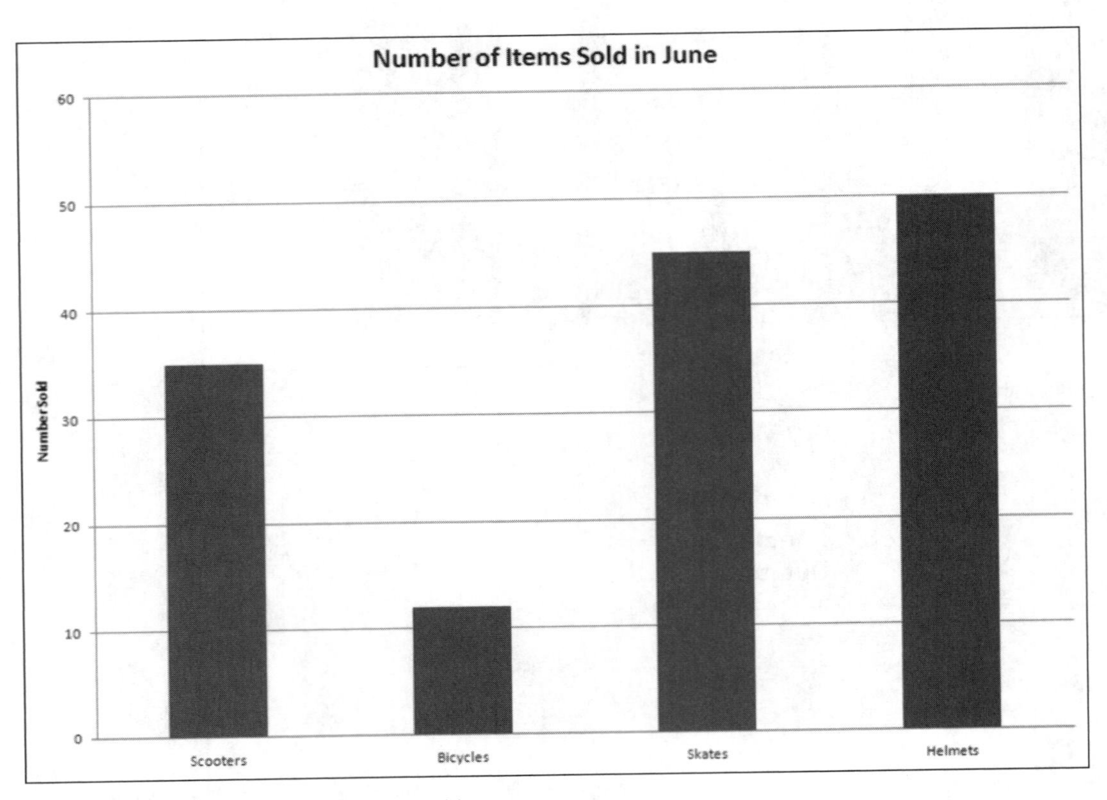

FIGURE F-7 Graph: number of items sold

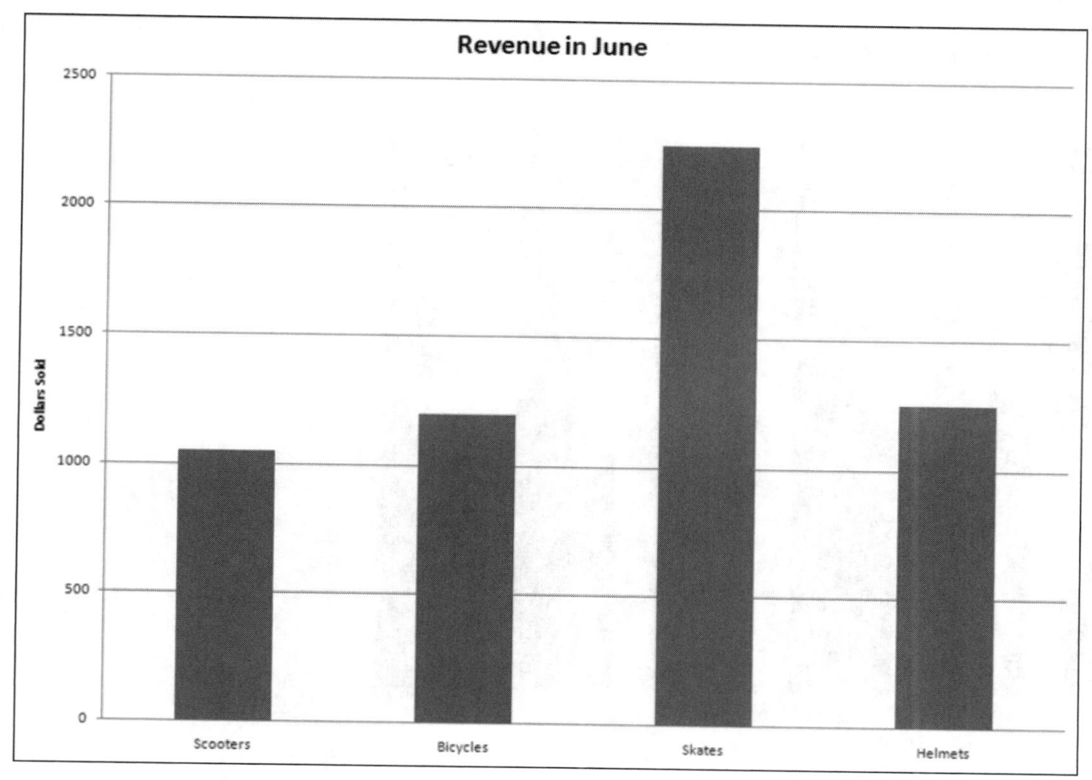

FIGURE F-8 Graph: profit of items sold

Something else you want to avoid is putting too much data in a single comparative chart. Here is an example: Assume you want to compare monthly mortgage payments for two loans with different interest rates and time frames. You have a spreadsheet that computes the payment data, shown in Figure F-9.

	A	B	C	D	E	F	G
1	**Calculation of Monthly Payment**						
2	Rate	6.00%	6.10%	6.20%	6.30%	6.40%	6.50%
3	Amount	$ 100,000	$ 100,000	$ 100,000	$ 100,000	$ 100,000	$ 100,000
4	Payment (360 payments)	$ 599	$ 605	$ 612	$ 618	$ 625	$ 632
5	Payment (180 payments)	$ 843	$ 849	$ 854	$ 860	$ 865	$ 871
6	Amount	$ 150,000	$ 150,000	$ 150,000	$ 150,000	$ 150,000	$ 150,000
7	Payment (360 payments)	$ 899	$ 908	$ 918	$ 928	$ 938	$ 948
8	Payment (180 payments)	$ 1,265	$ 1,273	$ 1,282	$ 1,290	$ 1,298	$ 1,306

FIGURE F-9 Calculation of monthly payment

In Excel, it is possible to capture all of that information in a single chart, such as the one shown in Figure F-10.

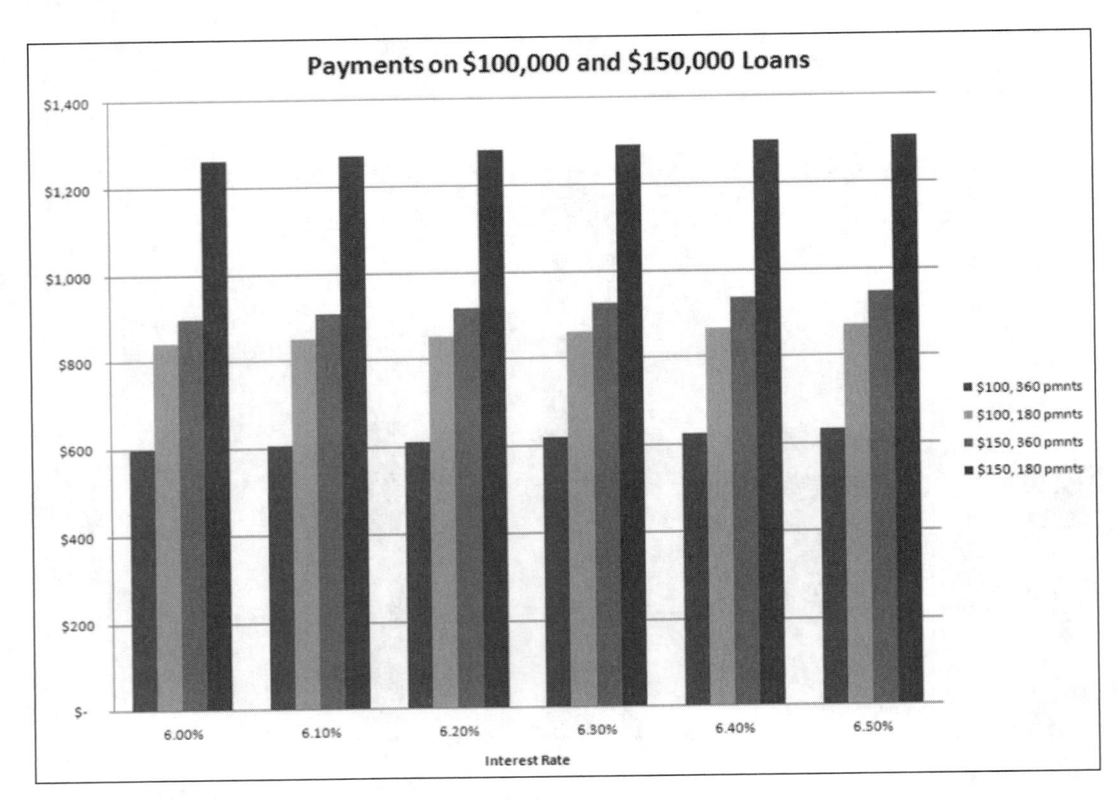

FIGURE F-10 Too much information in one chart

Note, however, that the chart contains a great deal of information. Most readers would probably appreciate your breaking down the information. For example, they would probably find the data easier to understand if you made one chart for the $100,000 loan and another chart for the $150,000 loan. The chart for the $100,000 loan would look like the one shown in Figure F-11.

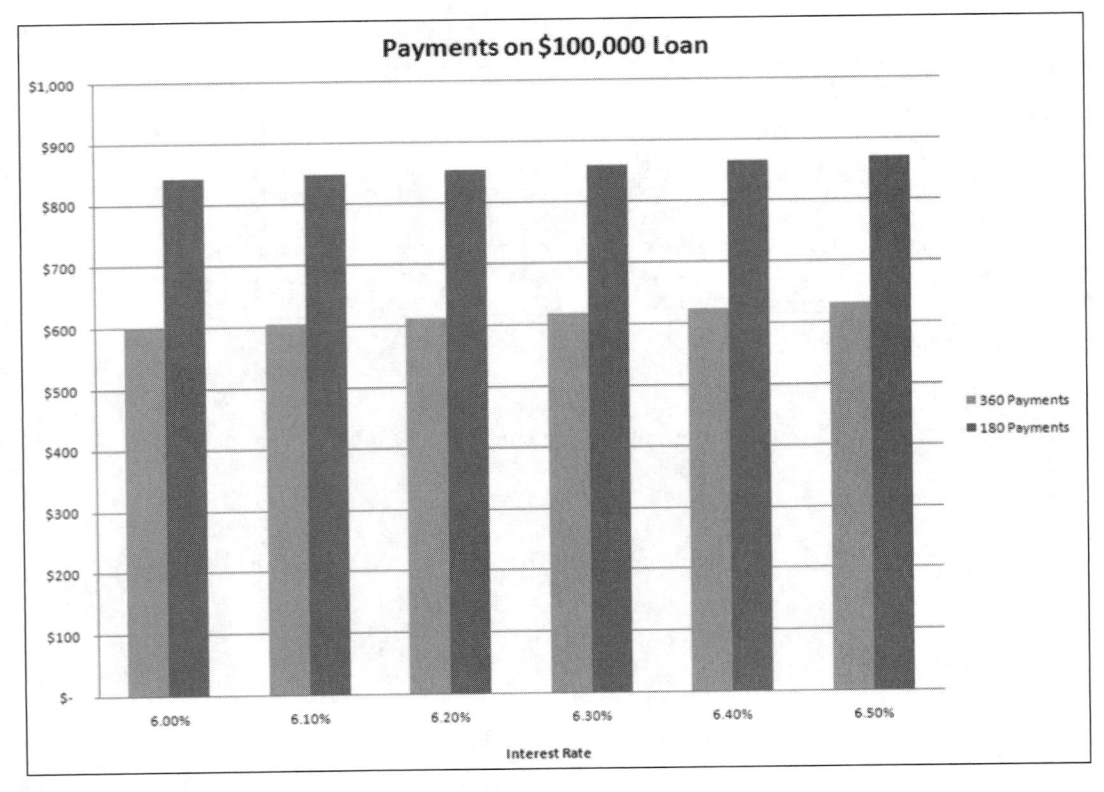

FIGURE F-11 Good balance of information

A similar chart could be made for the $150,000 loan. The charts could then be augmented by text that summarizes the main differences between the payments for each loan. In that fashion, the reader is led step-by-step through the data analysis.

You might want to use the Chart Wizard in Excel, but be aware that the charting functions can be tricky to use, especially with sophisticated charting. Some tweaking of the resultant chart is often necessary. Your instructor might be able to provide specific directions for your individual charts.

Creating PowerPoint Presentations

PowerPoint presentations are easy to create. Simply open the application and use the appropriate slide layout for a title slide, a slide containing a bulleted list, a picture, a graphic, and so on. When choosing a design template (the background color, the font color and size, and the fill-in colors for all slides in your presentation), keep these guidelines in mind:

- Avoid using pastel background colors. Dark backgrounds such as blue, black, and purple work well on overhead projection systems.
- If your projection area is small or your audience is large, consider using boldfaced type for all of your text to make it more visible.
- Use transition slides to keep your talk lively. A variety of styles are available for use in PowerPoint. Common transitions include dissolves and wipes. Avoid wild transitions such as swirling letters; they will distract your audience from your presentation.
- Use Custom Animation effects when you do not want your audience to see the entire slide all at once. When you use an Entrance effect on each bullet point on a slide, the bullets come up one at a time when you click the mouse or the right arrow. This Custom Animation effect lets you control the visual aid and explain the elements as you go along. These types of Custom Animation effects can be incorporated and managed under the Custom Animation screen, as shown in Figure F-12.

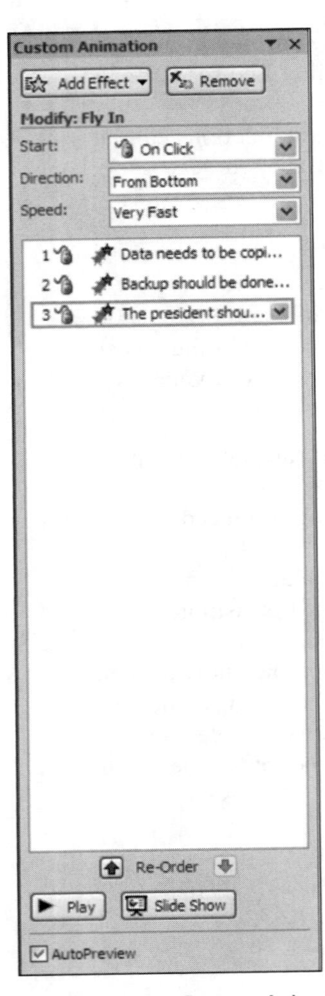

FIGURE F-12 Custom Animation screen

- Consider creating PowerPoint slides that have a section for your notes. You can print the notes from the Print dialog box by choosing Notes Pages from the Print what drop-down menu, as shown in Figure F-13. Each slide is printed half-size, with your notes appearing underneath each slide, as shown in Figure F-14.

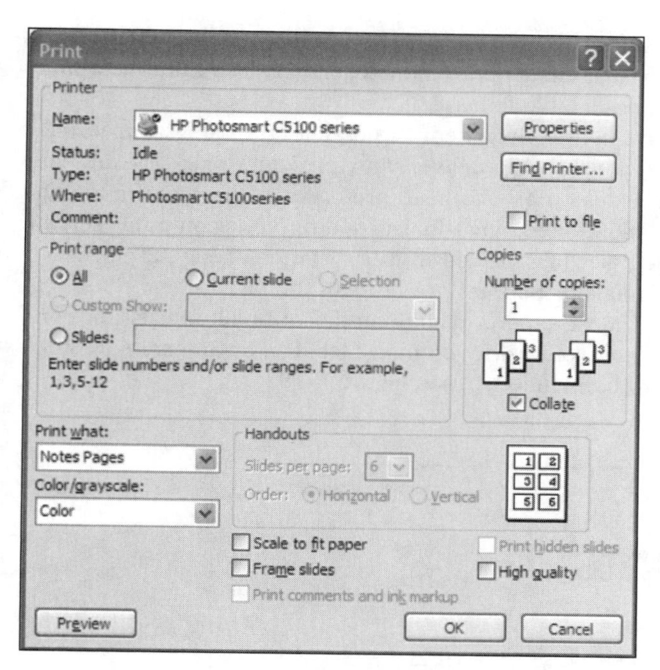

FIGURE F-13 Printing notes page

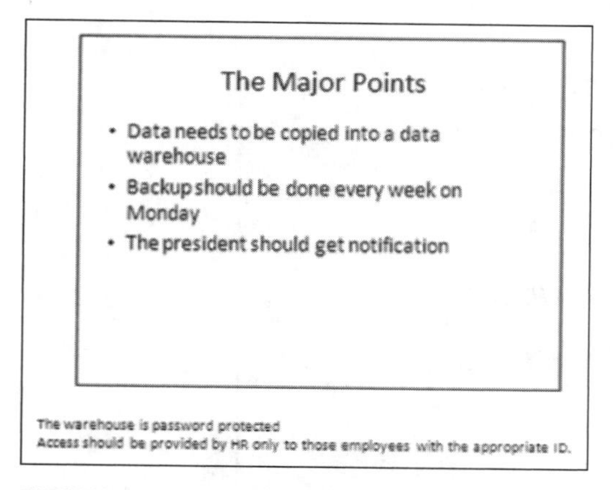

FIGURE F-14 Sample notes page

- You should check your presentation on an overhead. What looks good on your computer screen might not be readable on an overhead screen.

Using Visual Aids Effectively

Make sure you've chosen the visual aids that will work most effectively. Also make sure you have enough—but not too many—visual aids. How many is too many? The amount of time you have to speak will determine the number of visual aids you should use, as will your audience. For example, if you will be addressing a group of teenage summer helpers, you might want to use more visual effects than if you were making a presentation to a board of directors. Remember to use visual aids to enhance your talk, not to replace it.

Review each visual aid you've created to make sure it meets the following criteria:

- The size of the visual aid is large enough so that everyone in the audience can see it clearly and read any labels.
- The visual aid is accurate; for example, the graphics are not misleading, and there are no typos or misspelled words.
- The content of the visual aid is relevant to the key points of your presentation.

- The visual aid doesn't distract the audience from your message. Often when creating PowerPoint slides, speakers get carried away with the visual effects; for example, they use spiraling text and other jarring effects. Keep your visuals professional-looking.
- A visual aid should look good in the presentation environment. If possible, try using your visual aid in the presentation environment before you actually deliver your presentation. For example, if you'll be using PowerPoint, try out your slides on an overhead projector in the room in which you'll be showing the slides. As previously mentioned, what looks good on your computer screen might not look good on the overhead projector when viewed from a distance of, say, 20 feet.
- All numbers should be rounded unless decimals or pennies are crucial.
- Slides should not look too busy or crowded. Most experts recommend that bulleted lists not contain more than four or five lines. You want to avoid using too many labels. For an example of a slide that is too busy and therefore likely to be ineffective, see Figure F-15.

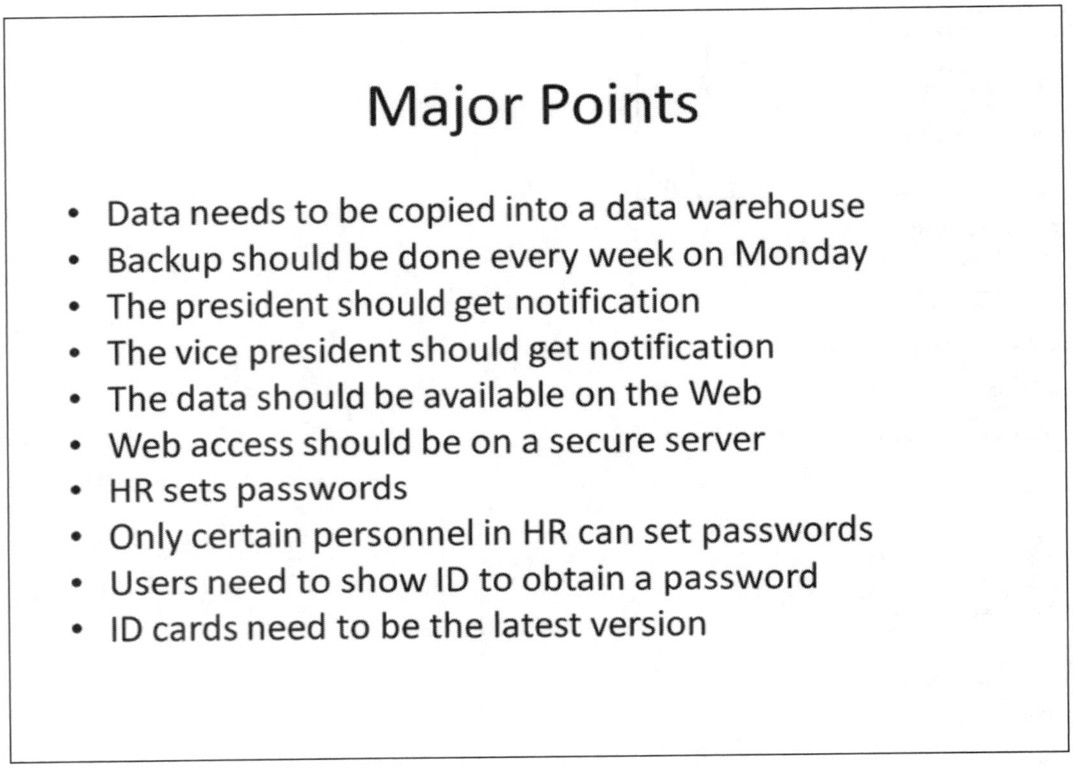

Major Points

- Data needs to be copied into a data warehouse
- Backup should be done every week on Monday
- The president should get notification
- The vice president should get notification
- The data should be available on the Web
- Web access should be on a secure server
- HR sets passwords
- Only certain personnel in HR can set passwords
- Users need to show ID to obtain a password
- ID cards need to be the latest version

FIGURE F-15 Busy slide

PRACTICING YOUR DELIVERY

Surveys indicate that for most people, public speaking is their greatest fear. However, fear or nervousness can be a positive factor. It can channel your energy into doing a good job. Remember that an audience is not likely to perceive you as being nervous unless you fidget or your voice cracks. Audience members want to hear the content of your talk, so think about them and their interests—not about how you feel.

The presentations you give for the cases in this textbook will be in a classroom setting with 20 to 40 students. Ask yourself this question: Am I afraid when I talk to just one or two of my classmates? The answer is probably no. Therefore, you should think of your presentation as an extended conversation with several of your classmates. Let your gaze shift from person to person, and make eye contact with various people. As your gaze drifts around the room, say to yourself: I'm speaking to one person. As you become more experienced in speaking before a group, you will be able to let your gaze move naturally from one audience member to another.

Tips for Practicing Your Delivery

Giving an effective presentation is not—and should not—be like reading a report to an audience. Rather, it requires that you rehearse your message well enough so you can present it naturally and confidently, in tandem with well-chosen visual aids. Therefore, you must allow sufficient time to practice your delivery before you give your presentation. Here are some tips to help you hone the effectiveness of your delivery:

- Practice your presentation several times, and use your visual aids when you practice.
- Show visual aids at the right time and only at the right time. A visual aid should not be shown too soon or too late. In your speaker's notes, you might include cues for when to show each visual aid.
- Maintain eye and voice contact with the audience when using the visual aid. Don't look at the screen or turn your back on the audience.
- Refer to your visual aids in your talk and with hand gestures. Don't ignore your own visual aid.
- Keep in mind that your visual aids should support your presentation, not *be* the presentation. In other words, don't include everything you are going to say on each slide. Use visual aids to illustrate key points and statistics, and fill in the rest of the content with your talk.
- Check your time. Are you within the time limit?
- Use numbers effectively. When speaking, use rounded numbers; otherwise, you'll sound like a computer. Also make numbers as meaningful as possible. For example, instead of saying "in 84.7 percent of cases," say, "in five out of six cases."
- Don't "reach" to interpret the output of statistical modeling. For example, suppose you have input many variables into an Excel model. You might be able to point out a trend, but you might not be able to say with certainty that if a company employs the inputs in the same combination, the firm will get the same results.
- Record and then evaluate yourself. If that is not possible, have a friend listen to you and evaluate your style. Are you speaking down to your audience? Is your voice unnaturally high-pitched from fear? Are you speaking clearly and distinctly? Is your voice free of distractions, such as *um*, *you know*, *uh*, *so*, and *well*?
- If you use a pointer, either a laser pointer or a wand, be careful. Make sure you don't accidentally direct a laser pointer toward someone's face—you'll temporarily blind the person. If you're using a wand, don't swing it around or play with it.

Handling Questions

Fielding questions from an audience can be an unpredictable experience, because you can't anticipate all of the questions you might be asked. When answering questions from an audience, *treat everyone with courtesy and respect*. Use the following strategies to handle questions:

- Try to anticipate as many questions as possible and prepare answers in advance. Remember that you can gather much of the information to prepare those answers while you draft your presentation. Also, if you have a slide that illustrates a key point but doesn't quite fit in your presentation, save it; someone might have a question that the slide will answer.
- Mention at the beginning of the talk that you will take questions at the end of your presentation. That should prevent people from interrupting you. If someone tries to interrupt, smile and say that you'll be happy to answer all questions when you're finished, or that the next graphic will answer the question. If, however, the person doing the interrupting is the CEO of your company, you should answer the question on the spot.
- When answering a question, repeat the question if you have *any* doubt that the entire audience has heard it. Then deliver the answer to the whole audience, not just the person who asked the question.
- Strive to be informative, not persuasive. In other words, use facts to answer questions. For instance, if someone asks your opinion about a given outcome, you might show an Excel slide that displays the Solver's output; then you can use that data as the basis for answering the question.
- If you don't know the answer to a question, don't try to fake it. For instance, suppose someone asks you a question about the Scenario Manager that you can't answer. Be honest. Say, "That is

an excellent question; but unfortunately, I don't know the answer." For the classroom presentations you will be delivering as part of this course, you might ask your instructor whether he or she can answer the question. In a professional setting, you can say that you'll research the answer and e-mail the results to the person who asked the question.

- Signal when you are finished. You might say, "I have time for one more question." Wrap up the talk yourself.

Handling a "Problem" Audience

A "problem" audience or a heckler is every presenter's nightmare. Fortunately, such experiences are rare. If someone is rude to you or challenges you in a hostile manner, remain cool, be professional, and rely on facts. Know that the rest of the audience sympathizes with your plight and admires your self-control.

The problem you will most likely encounter is a question from an audience member who lacks technical expertise. For instance, suppose you explained how to input data into an Access form but someone didn't understand your explanation. In that instance, ask the questioner what part of the explanation was confusing. If you can answer the question briefly, do so. If your answer to the questioner begins to turn into a time-consuming dialogue, offer to give the person a one-on-one explanation after the presentation.

Another common problem is someone who asks you a question that you've already answered. The best solution is to answer the question as briefly as possible, using different words (just in case the way in which you explained something confused the person). If the person persists in asking questions that have obvious answers, the person either is clueless or is trying to heckle you. In that case, you might ask the audience, "Who in the audience would like to answer that question?" The person asking the question should get the hint.

PRESENTATION TOOLKIT

You can use the form in Figure F-16 for preparation, the form in Figure F-17 for evaluation of Access presentations, and the form in Figure F-18 for evaluation of Excel presentations.

Preparation Checklist

Facilities and Equipment

- ☐ The room contains the equipment that I need.
- ☐ The equipment works and I've tested it with my visual aids.
- ☐ Outlets and electrical cords are available and sufficient.
- ☐ All the chairs are aligned so that everyone can see me and hear me.
- ☐ Everyone will be able to see my visual aids.
- ☐ The lights can be dimmed when/if needed.
- ☐ Sufficient light will be available so I can read my notes when the lights are dimmed.

Presentation Materials

- ☐ My notes are available, and I can read them while standing up.
- ☐ My visual aids are assembled in the order that I'll use them.
- ☐ A laser pointer or a wand will be available if needed.

Self

- ☐ I've practiced my delivery.
- ☐ I am comfortable with my presentation and visual aids.
- ☐ I am prepared to answer questions.
- ☐ I can dress appropriately for the situation.

FIGURE F-16 Preparation checklist

Evaluating Access Presentations

Course: _______ **Speaker:** _______________________ **Date:** ______

Rate the presentation by these criteria:
4=Outstanding 3=Good 2=Adequate 1=Needs Improvement
N/A=Not Applicable

Content

_____ The presentation contained a brief and effective introduction.
_____ Main ideas were easy to follow and understand.
_____ Explanation of database design was clear and logical.
_____ Explanation of using the form was easy to understand.
_____ Explanation of running the queries and their output was clear.
_____ Explanation of the report was clear, logical, and useful.
_____ Additional recommendations for database use were helpful.
_____ Visuals were appropriate for the audience and the task.
_____ Visuals were understandable, visible, and correct.
_____ The conclusion was satisfying and gave a sense of closure.

Delivery

_____ Was poised, confident, and in control of the audience
_____ Made eye contact
_____ Spoke clearly, distinctly, and naturally
_____ Avoided using slang and poor grammar
_____ Avoided distracting mannerisms
_____ Employed natural gestures
_____ Used visual aids with ease
_____ Was courteous and professional when answering questions
_____ Did not exceed time limit

Submitted by: ____________________________

FIGURE F-17 Form for evaluation of Access presentations

Evaluating Excel Presentations

Course: _______ **Speaker:** _______________________ **Date:** ______

Rate the presentation by these criteria:
4=Outstanding 3=Good 2=Adequate 1=Needs Improvement
N/A=Not Applicable

Content

_____ The presentation contained a brief and effective introduction.
_____ The explanation of assumptions and goals was clear and logical.
_____ The explanation of software output was logically organized.
_____ The explanation of software output was thorough.
_____ Effective transitions linked main ideas.
_____ Solid facts supported final recommendations.
_____ Visuals were appropriate for the audience and the task.
_____ Visuals were understandable, visible, and correct.
_____ The conclusion was satisfying and gave a sense of closure.

Delivery

_____ Was poised, confident, and in control of the audience
_____ Made eye contact
_____ Spoke clearly, distinctly, and naturally
_____ Avoided using slang and poor grammar
_____ Avoided distracting mannerisms
_____ Employed natural gestures
_____ Used visual aids with ease
_____ Was courteous and professional when answering questions
_____ Did not exceed time limit

Submitted by: ____________________________

FIGURE F-18 Form for evaluation of Excel presentations

INDEX

Note: **Bold** page numbers indicate where a key term is defined in the text.

E

F

G

H

I

J

K

L